Writes of Passage

Edward Said's oft-cited claim that Orientalists past and present have spun imaginative geographies where they sought ground truth has launched a plethora of studies of fictive geographies. Representations often reveal more about the culture of the author than that of the people and places represented. The study of imaginary geographies has raised many questions about Western writers' abilities to provide representations of foreign places; there is now much interest in the ways in which their productions of imaginative geographies reveal an intricate interplay between power, desire and place.

Writes of Passage explores the interplay between a system of 'othering', which travellers bring to a place, and the geographical differences they discover upon arrival. Exposing the tensions between the imaginary and the real, Duncan and Gregory and a team of leading international contributors focus primarily upon eighteenth- and nineteenth-century travellers to pin down the imaginary within the context of imperial power. The contributors focus on travel to three main regions: Africa, South Asia and Europe, with the European examples being drawn from Britain, France and Greece.

This book presents a unique contribution from geographers – with their sensitivity to issues of place, space and landscape – to contemporary studies on travel writing, migration and other related areas across geography, literary criticism, cultural studies and history. Offering important insights from leading writers in the field, as well as a critical survey of travel writing across an international spectrum, *Writes of Passage* represents a valuable addition to the burgeoning literature on travel writing.

Writes of Passage
Reading travel writing

Edited by
James Duncan and Derek Gregory

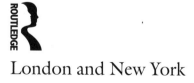

London and New York

First published 1999
by Routledge
11 New Fetter Lane, London EC4P 4EE

Simultaneously published in the USA and Canada
by Routledge
29 West 35th Street, New York, NY 10001

Typeset in Galliard by RefineCatch Limited, Bungay, Suffolk
Printed and bound in Great Britain by
TJ International Ltd, Padstow, Cornwall

British Library Cataloguing in Publication Data
A catalogue record for this book is available from the British Library

Library of Congress Cataloging in Publication Data
Writes of passage : reading travel writing / [edited by] James S.
 Duncan and Derek Gregory.
     p.   cm.
   1. Travel writing.   2. Travel in literature.   3. Travel – 18th
century.   4. Travel – 19th century.   5. Voyages, Imaginary.
I. Duncan, James S.   II. Gregory, Derek, 1951–   .
G151.W75   1999
808′.06691– dc21   98–8225

ISBN 0–415–16013–8 (hbk)
ISBN 0–415–16014–6 (pbk)

Contents

List of contributors   vii

1 **Introduction**   1
  JAMES DUNCAN AND DEREK GREGORY

2 **Limited Visions of Africa: Geographies of savagery and civility
  in early eighteenth-century narratives**   14
  ROXANN WHEELER

3 **Enlightenment Travels: The making of epiphany in Tibet**   49
  LAURIE HOVELL MCMILLIN

4 **Writing Travel and Mapping Sexuality: Richard Burton's
  Sotadic Zone**   70
  RICHARD PHILLIPS

5 ***The Flight from Lucknow*: British women travelling and writing
  home, 1857–8**   92
  ALISON BLUNT

6 **Scripting Egypt: Orientalism and the cultures of travel**   114
  DEREK GREGORY

7 **Dis-Orientation: On the shock of the familiar in a far-away place**   151
  JAMES DUNCAN

8 **The Exoticism of the Familiar and the Familiarity of the
  Exotic: *Fin-de-siècle* travellers to Greece**   164
  ROBERT SHANNAN PECKHAM

9 **Travelling through the Closet**   185
  MICHAEL BROWN

10 Writing Over the Map of Provence: The touristic therapy of
    A Year in Provence   **200**
    JOANNE P. SHARP

    Index   219

List of Contributors

Alison Blunt, Department of Geography, University of Southampton, Highfield, Southampton SO17 1BJ.

Michael Brown, Department of Geography, University of Washington, Seattle, Washington 98195.

James Duncan, Department of Geography, University of Cambridge, Downing Place, Cambridge CB2 3EN.

Derek Gregory, Department of Geography, University of British Columbia, Vancouver, BC V6T 1Z2.

Laurie Hovell McMillin, Expository Writing Program, Oberlin College, King Building 139, 10 North Professor Street, Oberlin, Ohio 44074–1019.

Robert Shannan Peckham, St Catherine's College, University of Cambridge, Cambridge CB2 1RL.

Richard Phillips, Department of Geography, Institute of Earth Studies, University of Wales, Aberystwyth, Dyfed SY23 3DW.

Joanne P. Sharp, Department of Geography, Glasgow University, Glasgow G12 8QQ.

Roxann Wheeler, Department of English, Indiana University of Pennsylvania, Indiana, PA 15705.

1   Introduction

James Duncan and Derek Gregory

FIELDS OF INTEREST

The closing decades of the twentieth century have witnessed a double explosion of interest in travel writing. On the one side, bookstores whose travel shelves were once confined to atlases, guidebooks and maps – to nominally 'factual' and 'objective' accounts – now include sections devoted to personal, avowedly imaginative accounts of travel. The style varies dramatically (even in the same book): from lush and lyrical to comic and picaresque, evoking a nineteenth-century tradition of exploration, enacting the ironic stance of late twentieth-century postmodernism. Many critics agree that the work of Bruce Chatwin, Pico Iyer or Redmond O'Hanlon – to name just three prominent authors – maps not only new landscapes in a still markedly various world but also new spaces within contemporary literature. They have driven travel writing beyond itself – some reviewers claim that they have even re-invented travel writing, giving it both a new popularity and a new critical respectability – by their determination to press new possibilities of finding the terms for – of coming to terms with – other cultures and other natures.

This sense of re-imagining the world through its re-presentation, describing spiralling circles between home and away, here and there, and reworking the connective between 'travel' and 'writing' gives much of this work a decidedly critical edge. At its very best, it raises urgent questions about the politics of representation and spaces of transculturation, about the continuities between a colonial past and a supposedly post-colonial present, and about the ecological, economic and cultural implications of globalizing projects of modernity. It is in this spirit, for example, that Appiah (1997) draws attention to the radically unsettling quality of O'Hanlon's account of his journey with James Fenton, *Into the Heart of Borneo* (1984):

> The real secret of O'Hanlon's success is that he subverts the conventions of this [natural-history] genre of imperial travel-writing by refusing utterly to take himself seriously. The imperial travellers – the explorers and naturalists – announced the difficulties of their journeys in order to record their triumphs over them. What they saw with their omnivorous eyes, they named 'properly'

for the first time, in the grand Linnaean manner. And the human indigenes of the forests – the native fauna of the genus Homo Sapiens – could be interesting, loyal, helpful, brave, even noble (as well, of course, as savage and stupid). But they were not likely to find the bwana-sahib ridiculous; and if they did, it was clear that this was further evidence of the error of their ways. In O'Hanlon's world, however, the natives are always amused.

(Appiah 1997, 20)

Similarly, in his epic *Congo Journey* (1996) O'Hanlon sets out what Appiah takes to be 'some of the starker complexities that confront anyone from the richer world who enters a place of such intense material deprivation'. Much of this is familar ground, but it is worked over in a different way by O'Hanlon's travelling companions, Larry Shaffer (an American professor of psychology) and Marcellin Agnaga (a Congolese biologist). Appiah remarks on the chronic struggle between Shaffer's anti-relativism and Agnaga's insistence on the ineradicable specificity of Africa, and the brooding tension between Agnaga's modern, rationalist intellect and his anger at O'Hanlon 'for drawing him back into the ambit of his grandfather and his sorcery'. O'Hanlon offers no easy answers to these predicaments, still less to the troubling question posed by Appiah himself: 'What about the moral muddles of those of us who can only have access to this world through the accounts of travellers like O'Hanlon?'

What indeed? On the other side, then, and paralleling these literary sensibilities, has emerged a new academic interest in travel writing which often poses similarly disconcerting questions in its return to this decidedly nineteenth-century tradition of exploration and travel. This was hardly a *terra incognita* to traditional histories of geographical knowledge, to be sure, but travel writing that was not conducted under the sign of 'Science' was virtually ignored by such studies. What remained was read as an unproblematic record of heroism and triumphant discovery in which other cultures and other natures were shown to have surrendered their secrets before the powerful gaze of Western 'Reason'. In effect, these studies mapped the production of a space of knowledge but not the concomitant production of a space of power.

More recently, however, scholars have started to cast their nets into two much wider streams of work. In the first place, drawing on the thematics of Raymond Williams and others, travel and its cultural practices have been located within larger formations in which the inscriptions of power and privilege are made clearly visible. We are thus beginning to understand much more about the cultures of natural history, for example, and the complex dialectic between scientific expeditions in the field and the circulation of their knowledges through metropolitan and colonial centres of calculation (Jardine, Secord and Spary 1996; Miller and Reill 1996). Considerable attention has been paid to the ways in which extra-scientific travel entered into ideologies of nationalism, to take another example, and contributed to the consolidation of bourgeois cultures through the refinement of 'landscape tastes' (Andrews 1989; Buzard 1993; Ousby 1990; Pemble 1987). In the second place, in an approach usually more self-consciously theoretical in its inclinations and drawing on both poststructuralism and post-colonialism, travel writing has been analysed as an ensemble of textual practices that can be made to disclose the characteristic gestures of an 'imperial stylistics' (Pratt 1992; see also Spurr 1993). There is a sense in which all travel writing, as a process of inscription and appropriation, spins webs of colonizing power, but to locate travel writing within this discursive formation also involves plotting the play of fantasy and desire, and the possibility of transgression (Aldrich 1993; Barrell 1991; Porter 1991). Particular attention has been paid to the articulations of travel, gender and sexuality and the routes by which travel writings modulated and registered the changing constitution of travelling subjects; and here the masculinism of traditional histories of 'exploration' has yielded to a principled recovery of the complex subject-positions of both men and women travellers (Blunt 1995; Fawley 1994; Lawrence 1994; Melman 1992; Mills 1991).

We welcome these developments – and the connections between them – but we want to make several cautionary observations.

SPACES OF REPRESENTATION

We think it important to register the physicality of representation itself. This involves attending to the multiple sites at which travel writing takes place and hence to the spatiality of representation. We know, for example, that the published record of Captain Cook's third voyage is a composite account. The 'official' version, prepared for public circulation by the Admiralty, was the fourth and final stage of a sequence that began with the original entries in the ship's log, passed through Cook's own journal, and these observations then fed into a manuscript that was revised and edited by John Douglas. As the sequence proceeds so the writing is progressively distanced from the events and scenes it purports to convey (MacClaren 1992). More than this: in a brilliant re-reading of Cook's arrival at Nootka Sound, Clayton (1998a; 1998b) shows that even when his ships were stationary, Cook's journal 'was still on the move'. He constantly reworked his observations in the light of subsequent encounters, so that his path across the Pacific to the west coast of North America produced not only a space of observation but a space of comparison, in which each journal entry was mediated and modified by its shifting articulations with other entries in the series (cf. Carter 1987). Besides, Cook's officers kept their own journals, and they often saw different things from different positions and had different encounters with native peoples. These journals were collected by the Admiralty and incorporated into the official account so that, as Clayton says, 'truth and objectivity became still more decentred'. There can be no question of Douglas's authorized version capturing any presumptive immediacy and transparency of the event: it is a composite, fractured and spatialized construction. When Flaubert travelled up the Nile in 1849–50, to take an example from a different cultural register, he kept a journal which he used not only as an *aide-mémoire* but also as the basis for letters to his mother and to his male friends at home in France (so that there are often different constructions of the same event or place). On his return Flaubert collected the letters and used them, in conjunction with his own journal, to compose yet another account of his *Voyage en Égypte* (Gregory 1995). Here too the text is a composite, and it folds into itself a series of different spatialities that entered into its own construction.

More than the movements of the pen, it is also necessary to attend to the different means by which travellers recorded their experiences. Thus Cook was accompanied on his first Pacific voyage by both draftsmen and artists, and Smith (1985) has shown that there was a complex dialogue between the contrasting traditions that they brought to bear on their work. It is not simply that they brought with them to the South Seas the cultural baggage of late eighteenth-century Europe: it is a question of recognizing the different integrities and conventions of the various media that were available to explorers and travellers, and hence of triangulating the space of representation (see also Stafford 1984). Or again, to return to our second example, Flaubert was accompanied up the Nile by Maxime Du Camp, who took with him one of the first photographic apparatuses to be used in Egypt. We know from their writings that the two friends had different preoccupations: that du Camp was obsessed by photographing the temples and tombs of ancient Egypt whereas Flaubert was frankly bored by the whole business and much more interested in the physicality and sensuality of nineteenth-century Egypt (Gregory 1995). But when du Camp's photographic record is inserted into these textual compositions it maps no simple 'ground truth' but, once again, a complex and foliated space of representation whose critical interpretation has much to tell us about the material production of imaginative geographies (MacClaren 1992). All of this suggests a second sense in which the travel archive is fractured. Too often, we think, journals, letters and published writings are assigned to literary scholars and historians; sketches, water-colours and paintings to art historians; and photographs and postcards to historians of photography. We suggest that the alternative strategy of attending to the physicality of representation imposes the obligation to read these different media together and, in so doing, to attend to their different valences and silences.

There is yet a third sense in which we want to accentuate the spatiality of representation: travel writing as an act of translation that constantly works to produce a tense 'space in-between'. Defined literally, 'translation' means to be transported from one place to another, so that it is caught up in a complex dialectic between the recognition and recuperation of difference (Miller 1996). Memory, especially collective or social memory, is also a form of translation 'marked by a boundary crossing and by a realignment of what has become different' (Iser 1996, 297; see also Motzkin 1996, 265–81). But, as Maurice Halbwachs's study of the cultural construction of 'the Holy Land' and its pilgrimage routes reminds us, social memory is often sedimented through circuits in space. In re-presenting other cultures and other natures, then, travel writers 'translate' one place into another, and in doing so constantly rub against the hubris that their own language-game contains the concepts necessary to represent another language-game (Dingwaney 1995, 5; Asad and Dixon 1973; 1985). Just as textual translation cannot capture all of the symbolic connotations of language or the alliterative sound of words, the translation of one place into the cultural idiom of another loses some of the symbolic loading of the place for its inhabitants and replaces it with other symbolic values. This means that translation entails both losses and gains, and as descriptions move from one place to another so they circulate in what we have called 'a space in-between'. This space of translation is not a neutral surface and it is never innocent: it is shot through with relations of power and of desire. In general, and as Venuti (1993, 210) points out, translation is either a 'domesticating method, an ethnographic reduction of the foreign text to target language cultural values, bringing the author back home' or a 'foreignizing method, an ethnographic pressure on those values to register the linguistic and cultural difference of the foreign text, sending the reader abroad'. Travel writing is often inherently domesticating, in something like Venuti's sense, and we have noted how many critics have emphasized its complicity with the play of colonizing power. But even in its most imperial gestures, by virtue of its occupation of that 'space in-between' – the space of transculturation (Pratt 1992) – travel writing can also disclose an ambivalence, a sense of its own authorities and assumptions being called into question.

SPACES OF TRAVEL

It may seem strange to emphasize the spatiality of travel, but we fear that some critical readings fail to register the production of travel writings by corporeal subjects moving through material landscapes. Such critiques are vulnerable to the accusation of textualism: they tell us much about the constitution of authors as subjects through the process of writing, and about the relations between their own strategies of representation and the wider cultural formations of which they were a part, but they often say very little about the places these travellers encountered or the physical means through which they engaged them. This is not to call for any comparison between 'imaginative' geographies and 'real' geographies, as persistent misreadings of Edward Said's critique of Orientalism would have us believe, because all geographies are imaginative geographies – fabrications in the literal sense of 'something made' – and our access to the world is always made through particular technologies of representation.

Our focus in this collection is on European and American travellers writing between the eighteenth and twentieth centuries. It was during this period that travel assumed a characteristically modern form. By this, we mean that travel and travel writing entered into the Euro-American project of modernity in at least three ways.

First, travel and travel writing meshed with secularization: sacralized frames of reference yielded to a much more complex taxonomy of cultural difference and natural history. Within the geographical imaginary of post-Enlightenment Europe, 'Asia', 'Africa', 'America' and 'Australia' were each discursively constituted by relations of contradiction and opposition that not only confirmed

'Europe' as sovereign subject but also marked out a differentiated and often agonistic space of alterity (Gregory 1998). And while there was a systematicity, a labile coherence, to this imaginary, it was constituted as a comparative matrix at several different levels and there was no single space of representation to be occupied by unitary descriptions of, say, 'India', or 'the Orient', or even 'Africa' (Teltscher 1995; Lowe 1991; Youngs 1994).

Second, travel became more than a necessary evil, a burden to be borne by, for example, pilgrims, merchants and explorers, but rather came to be constructed as an end in itself, as a form of pure pleasure. To be sure, the aristocratic Grand Tour had been an uneasy enterprise since its inception in the sixteenth century (Black 1992): there was a decided ambiguity about the exposure of impressionable young men to cultural difference even within Europe (Warneke 1995), and young women were widely assumed to be compromised by the very idea (Richter 1995; Craik 1997, 119). By the end of the eighteenth century, however, the fragile compromise implicit in the Grand Tour's vaunted promise of what Batten (1978) calls 'pleasurable instruction' had begun to break down: 'pleasure overwhelm[ed] the demand for instruction' (Porter 1991, 125). And yet, as Porter goes on to demonstrate, even this was shot through with ambivalence and contradiction. If travel allowed for spaces of licence and liminality, these were often bounded: modern travellers made their forays into an exotic, sometimes perverse and on occasion thrilling world from the security of vantage points and viewing platforms that were recognizably their own, and their practices and performances were scripted by travel writers and guidebooks.

Third, travel was no longer an exclusively aristocratic preserve. To be sure, it continued to be privileged in all sorts of ways, but in the course of the nineteenth century it was increasingly construed as a quintessentially bourgeois experience that had its origins in the conjunction of romanticism and industrialism. Romanticism marked a post-Enlightenment remapping of the space of representation: it de-throned the sovereignty of Reason and glorified unconstrained impulse, individual expression and the creative spirit. It celebrated what Cardinal (1997) calls the 'agitations of personal perception' over objective observation. The first stage of romantic travel began in the late eighteenth century, and after the hiatus of the Napoleonic wars carried on more or less unchallenged until at least the mid-nineteenth century. Central to romantic travel was a passion for the wildness of nature, cultural difference and the desire to be immersed in local colour. Such travel was seen as most likely to accomplish its goals if it was slow, unregimented and solitary. By the nineteenth century its most characteristic figure was the young bourgeois fleeing the dull repetitions and the stifling mundanity of the bourgeois world: tourism, says Hans Magnus Enzensburger (1996, 125), 'is thus nothing other than the attempt to realize the dream that Romanticism projected onto the distant and the far away'. And yet the terrible irony was that modernity was already there, lying in wait. Enzensburger sees this as the central dialectic of modern tourism: 'Tourism is always outrun by its refutation . . . [yet] far from resigning and giving up the struggle at the cost of freedom, tourism redoubles its efforts after each defeat.'

The dilemma was double-edged. On the one side, most modern travellers prepared for their journeys by reading the accounts of other travellers and noting the recommendations of the guidebooks. If this helped to reduce the anxiety of travel then it also aggravated what Porter (1991, 12) has called 'the anxiety of travel-writing': in writing their own accounts of the beaten track, travellers had to find some way of describing the familiar in novel and entertaining ways (Buzard 1993). In effect, the citationary structure of these accounts gave modern travel a routinized and repetitive form and yet this threatened the very integrity and 'authenticity' of the experience itself. On the other side, the networks that made escape possible – railways and steamships, hotels and tour companies – ensured that modern tourism was constantly haunted by the spectre of belatedness, by the sense of arriving at the very moment that a non-modern world was fast disappearing under the impress of modernity (Behdad 1994; see also Terdiman 1985). This crisis was exacerbated during the second stage of romantic travel not only by the obvious signs of colonial occupation and capital accumulation in the landscape – barracks and ministries, docks and railways, factories and plantations – but by the presence of other travellers. Thomas Cook's first tour in 1851 marked the extension of modern travel to the petty bourgeoisie and fractions of the working class (Brendon 1991; Withey 1997). They too sought a romantic experience, but one that was well ordered and regimented – a sort of 'industrialized' romanticism – and the old élites sharpened their pens against the infringements and invasions of their own hitherto exclusive spaces of travel by common 'tourists' (see Buzard 1993).

As the nineteenth century wore on, so the sheer number of tourists present in some places made the illusion of discovery, or even immersion in the local, harder to sustain. The romantic's idea of an unsullied world was replaced by the image of a world overrun by industrialism. For some romantics the world had been so spoiled that it was best to travel only in one's imagination. In his *Voyage en Orient* (1857) Nerval writes:

> O my dear friend, how perfectly we have enacted the fable of the two men, of whom one scurries to the ends of the earth in search of his good fortune, while the other quietly awaits it in his own domestic bed! . . . Only once, out of imprudency, did you change your idea of Spain by going to see it. . . . But already I have lost kingdom by kingdom, province by province, the most beautiful half of the universe, and soon I shall no longer know where to seek a refuge for my dreams.
>
> (quoted in Cardinal 1997, 151)

At the end of the twentieth century we are still in the age of 'industrialized' romanticism. By this we mean that although the bureaucratization of travel has increased since the turn of the century, the romantic frame through which places are viewed remains much the same. Although post-colonial theorists write of 'travelling cultures' (Clifford 1992; 1997) and travelling theory (Behdad 1994; Grewal and Kaplan 1994; Kaplan 1996; Robertson *et al.* 1994; Rojek and Urry

1997), travel is still popularly understood as the immersion in picturesque, distinct, colourful cultures. Furthermore, travel to the former colonies is still scripted in a nineteenth-century romantic mind set characterized by Said (1979; 1993) as Orientalist, and by Fabian (1983) as a form of time–space substitution. The romantic longing for a world that has been lost is expressed in the late twentieth century as heritage in Europe and as an 'imperialist nostalgia' (Rosaldo 1989) in the former colonies, or what Behdad (1994) has termed a sense of loss experienced as belatedness.

AN INTRODUCTION TO THE ESSAYS

Roxann Wheeler explores how certain eighteenth-century travel texts focus on what Laura Brown (1993) has termed a 'rhetoric of acquisition', a cataloguing of commodities and resources, rather than a rhetoric of the aesthetic. Also largely absent from these texts is another important nineteenth-century 'scientific'/ aesthetic category, blackness. By examining early eyewitness accounts of Africa and the newly emerging forms of the novel and natural history, Wheeler demonstrates that even when race *was* considered, it was constructed largely without reference to skin colour. For example, Daniel Defoe's *Captain Singleton* (1720) does not consider colour noteworthy. Although Africans' skin colour was sometimes described by British travellers, its later status as a sign of inferiority was not well established in the 1720s when the novel was published. Much more significant racialized categories included savagery and nakedness. We can see that what seemed to later commentators to be one of the most noteworthy, empirically obvious marks of difference was so only within an historically and culturally specific discursive frame.

While the texts Roxann Wheeler examines are firmly anchored within an Enlightenment frame, the travel texts of Tibet, on which Laurie McMillin concentrates, span the shift from the Enlightenment to early twentieth-century romantic modes of writing. She traces the development of the myth that travel in Tibet produces personal transformation. She shows that this often self-fulfilling myth is part of a larger British project to make sense of selves and others during the colonial era. Having said this, McMillin argues that any analysis which attempts to understand the operation of general colonial discourses or imperialist ideologies misses the specificity of a place. She argues for instance that the specificity of the construction of Tibet arises out of the unique conjuncture of high mountains and Buddhist state religion with a British cult of the romanticism of the wild mountain landscape ideal.

Richard Phillips examines the mapping of Victorian sexuality through a detailed, contextual reading of the nineteenth-century explorer and travel writer Richard Burton. Burton mapped geographies of sexuality, among them a neatly bounded 'Sotadic Zone', in which he claimed pederasty was common. Mapped onto a precisely bounded geographical area, homosexual practices were then spatially displaced and racially essentialized. By corollary, British heterosexuality was constructed, simplified and morally purified. The sexualities Burton charts reflect the geography of his travels: a polarized geography, divided between home and away, colonizer and colonized, Britain and the 'Sotadic Zone', maps polarized, radicalized and essentialized sexualities. Discrete though these spaces and identities appear to be, Burton nevertheless travels through and between them, transgressing apparently immutable boundaries and betraying the instability of his own geographical and sexual constructions.

Burton's writings are much more characteristic of Enlightenment travel accounts than of romantic ones, in that his writings combine elements of exploration with a recuperation of cultural difference to allow moral commentary on home.

Alison Blunt examines imperial travel writing by women in terms of collective rather than individual travel, and under forced rather than voluntary conditions. She does so by focusing on women's writings in the three months after the siege of Lucknow in the Indian uprising of 1857–8. She argues that these accounts provide a lens on the ways in which imperial power and authority were reconstructed as these women travelled further away from the symbolically charged site of Lucknow. While Blunt argues that the writings of these women were highly spatialized, their writings dwell largely upon their inner states rather than a description of the countryside. And yet these were not the romanticized inner states of the romantic traveller. Once again we can discern both a complexity of imagination and situational specificity in the nature of travel accounts.

Derek Gregory examines the ambiguities and contradictions in the scriptings of Egypt by European and American travellers in the nineteenth and early twentieth centuries. By focusing on different types of romantic travellers, from the solitary traveller to the tourist on the Cook's tours, Gregory raises a series of questions about the politics of vision and the imaginative geographies inscribed through colonial sightseeing. In doing so he directly engages Edward Said and his critics on the issue of imaginative geographies. Gregory argues that imaginative geographies script a place and in so doing make the imaginative real, both materially and performatively.

James Duncan explores Homi Bhabha's notion that the civilizing mission of colonialism 'often produces a text rich in the traditions of trompe-l'oeil, irony, mimicry and repetition'. One site of this mimicry is the Kandyan Kingdom in the interior mountains of Ceylon, which was transformed in the nineteenth century into a region of plantations and tourist sites. This transformation included both the real and the imaginative reworkings of the landscape into a mirror of the Lake District of northern England. Deconstructing the accounts of British tourists and planters, Duncan demonstrates how this 'mimic place' had a dis-orienting effect, constituting not a landscape of pure alterity, but one which itself incorporated self/home and other/away.

Shannan Peckham argues that European travellers to Greece in the nineteenth century arrived with largely unchallenged assumptions about the centrality of ancient Greece to their own cultural histories. Greek places were mediated through a body of texts which have had (especially since the eighteenth century) the effect of shifting the origins of European culture from Rome to Greece. Contemporary Greece was another matter, however. At once familiar and exotic, the ambiguity of Greece's geographical and cultural position preoccupied Orientalist travellers in the nineteenth century. Peckham argues that the unstable geographical and cultural taxonomies within which European travellers sought to define Greece shed light on a latent anxiety about the place of European culture during a period of colonial expansion when 'exotic' cultures were becoming progressively familiar and vice versa. Greece, the author claims, greatly complicates the dualism of Orient and Occident which Edward Said posits in his Orientalist thesis.

Michael Brown explores the spaces of home and travel for gay men in the 1980s and 1990s. Adopting a psychoanalytic approach to desire, Brown examines the writings of Neil Miller for what they have to say about gay communities in different places and their attitudes towards sexuality. He employs the spatial metaphor of 'the closet' as a 'translation term' between sexuality, imagined space and desire. Miller's writings resonate with those of Burton one hundred years earlier, for both use travel writing as a basis for commenting on contemporary moral issues concerning the self.

Joanne Sharp examines Peter Mayle's *A Year in Provence*, one of the most popular travel accounts of recent years. However, the very success of this and his other travel books has made his Provence unobtainable. The imagined geography of the region that he produced came to obscure other images of Provence in the media. Ironically, the appeal of this image transformed the actual place, as the Provençal landscape became increasingly overrun by British visitors who had come to see Mayle's France. More recently there has been a backlash against both his writing and his part of Provence, as tourists seek out more 'authentic', less popularized parts of the region.

REFERENCES

Aldrich, Robert (1993) *The Seduction of the Mediterranean: Writing, art and homosexual fantasy*, London: Routledge.

Andrews, Malcolm (1989) *The Search for the Picturesque: Landscape aesthetics and tourism in Britain, 1760–1800*, Stanford: Stanford University Press.

Appiah, Anthony (1997) 'Mokélé-Mbembe, being the Faithful Account of a Hazardous Expedition to find the Living African Dinosaur', *London Review of Books*, 24 April, pp. 19–21.

Asad, T. and J. Dixon (1985) 'Translating Europe's others', in F. Barker *et al.* (eds), *Europe and its Others*, Colchester: University of Sussex, pp. 170–93.

Barrell, John (1991) 'Death on the Nile: fantasy and the literature of tourism, 1850–60', *Essays in Criticism* 41, pp. 97–127.

Batten, Charles (1978) *Pleasurable Instruction: Form and convention in eighteenth-century travel literature*, Berkeley: University of California Press.

Behdad, Ali (1994) *Belated Travelers: Orientalism in the age of colonial dissolution*, Durham, NC: Duke University Press.

Black, Jeremy (1992)*The British Abroad: The Grand Tour in the eighteenth century*, New York: St Martin's Press.

Blunt, Alison (1995) *Travel, Gender and Imperialism: Mary Kingsley and West Africa*, New York: Guilford Press.

Brendon, P. (1991) *Thomas Cook: 150 years of popular tourism*, London: Secker and Warburg.

Brown, L. (1993) *Ends of Empire: Women and ideology in early eighteenth-century English literature*, Ithaca: Cornell University Press.

Buzard, James (1993) *The Beaten Track: European tourism, literature and the ways to 'culture' 1800–1918*, Oxford: The Clarendon Press.

Cardinal, R. (1997) 'Romantic travel', in Roy Porter (ed.), *Rewriting the Self*, London: Routledge, pp. 135–55.

Carter, Paul (1987) *The Road to Botany Bay*, London: Faber.

Clayton, Daniel (1998a) 'Captain Cook and the spaces of contact at Nootka Sound', in Jennifer Brown and Elizabeth Vibert (eds), *Documenting Native History: Texts and contexts*, Peterborough, ON, and Calgary, AB: Broadview Press.

—— (1998b) *Islands of Truth*, Berkeley: University of California Press.

Clifford, J. (1992) 'Travelling cultures', in L. Grossberg, C. Nelson and P. Treichler (eds), *Cultural Studies*, London: Routledge, pp. 96–116.

—— (1997) *Routes: Travel and translation in the late twentieth century*, Cambridge, MA: Harvard University Press.

Craik, J. (1997) 'The culture of tourism', in C. Rojek and J. Urry (eds), *Touring Cultures: Transformations of travel and theory*, London: Routledge, pp. 113–36.

Dingwaney, A. (1995) 'Introduction: translating "third world" cultures', in A. Dingwaney and C. Maier (eds), *Between Languages and Cultures: Translation and cross-cultural texts*, Pittsburgh: Pittsburgh University Press, pp. 3–15.

Enzensburger, Hans Magnus (1996) 'A theory of tourism ', *New German Critique* 68, pp. 117–35 [originally published in *Merkur* 126 (1958), pp. 701–20].

Fabian, J. (1983) *Time and the Other*, New York: Columbia University Press.

Fawley, M. (1994) *A Wider Range: Travel writing by women in Victorian England*, Oxford: Oxford University Press.

Gregory, Derek (1995) 'Between the book and the lamp: imaginative geographies of Egypt, 1849–50', *Transactions of the Institute of British Geographers* 20, pp. 29–57.

—— (1998) 'Power, knowledge and geography', *Geographische Zeitschrift* 86.

Grewal, I. and Kaplan, C. (eds) (1994) *Scattered Hegemonies: Postmodernity and transnational feminist practice*, Minneapolis: University of Minnesota Press.

Iser, W. (1996) 'Coda to the discussion', in S. Budick and W. Iser (eds), *The Translatability of Cultures: Figurations of the space between*, Palo Alto: Stanford University Press, pp. 95–302.

Jardine, N., Secord, J.A. and Spary, E.C. (eds) (1996) *Cultures of Natural History*, Cambridge: Cambridge University Press.

Kaplan, C. (1996) *Questions of Travel: Postmodern discourses of displacement*, London: Duke.

Lawrence, Karen (1994) *Penelope Voyages: Women and travel in the British literary tradition*, Ithaca: Cornell University Press.

Lowe, Lisa (1991) *Critical Terrains: French and British Orientalisms*, Ithaca: Cornell University Press.

MacClaren, I.S. (1992) 'Exploration/travel literature and the evolution of the author', *International Journal of Canadian Studies* 5, pp. 39–68.

Melman, Billie (1992) *Women's Orients: English women and the Middle East, 1718–1918*, London: Macmillan.

Miller, David and Reill, Peter (eds) (1996) *Visions of Empire: Voyages, botany and representations of nature*, Cambridge: Cambridge University Press.

Miller, J.H. (1996) 'Border crossings, translating theory: Ruth', in S. Budick and W. Iser (eds), *The Translatability of Cultures: Figurations of the space between*, Palo Alto: Stanford University Press, pp. 207–23.

Mills, Sara (1991) *Discourses of Difference: An analysis of women's travel writing and colonialism*, London and New York: Routledge.

Mitchell, T. (1988) *Colonizing Egypt*, Berkeley: University of California Press.

Motzkin, G. (1996) 'Memory and cultural translation', in S. Budick and W. Iser (eds), *The Translatability of Cultures: Figurations of the space between*, Palo Alto: Stanford University Press, pp. 265–81.

O'Hanlon, R. (1984) *Into the Heart of Borneo*, London: Picador.

—— (1996) *Congo Journey*, London: Penguin.

Ousby, Ian (1990) *The Englishman's England: Travel, taste and the rise of tourism*, Cambridge: Cambridge University Press.

Pemble, John (1987) *The Mediterranean Passion: Victorians and Edwardians in the South*, Oxford: Oxford University Press.

Porter, Dennis (1991) *Haunted Journeys: Desire and transgression in European travel writing*, Princeton: Princeton University Press.

Pratt, Mary Louise (1992) *Imperial Eyes: Travel writing and transculturation*, London and New York: Routledge.

Richter, L. (1995) 'Gender and race: neglected variables in tourism research', in R. Butler and D. Pearce (eds), *Change in Tourism: People, places, processes*, London: Routledge, pp. 79–96.

Robertson, G. *et al.* (1994) *Travellers' Tales: Narratives of home and displacement*, London: Routledge.

Rojek, C. and Urry, J. (eds) (1997) *Touring Cultures: Transformations of travel and theory*, London: Routledge.

Rosaldo, R. (1989) *Truth and Culture: The remaking of social analysis*, Boston: Beacon.

Said, E. (1979) *Orientalism*, New York: Vintage.

—— (1993) *Culture and Imperialism*, New York: Knopf.

Smith, Bernard (1985) *European Vision and the South Pacific*, New Haven: Yale University Press.

Spurr, David (1993) *The Rhetoric of Empire: Colonial discourse in journalism, travel writing and imperial administration*, Durham, NC: Duke University Press.

Stafford, Barbara (1984) *Voyage into Substance: Art, science, nature and the illustrated travel account 1760–1840*, Cambridge, MA: MIT Press.

Teltscher, Kate (1995) *India Inscribed: European and British writing on India 1600–1800*, New Delhi: Oxford University Press.

Terdiman, Richard (1985) 'Ideological voyages: a Flaubertian dis-Orient-ation', in his *Discourse/Counter-discourse: The theory and practice of symbolic resistance in nineteenth-century France*, Ithaca: Cornell University Press, pp. 227–57.

Venuti, L. (1993) 'Translation as cultural politics: regimes of domestication in English', *Textual Practice* 7.2, pp. 208–23.

Warneke, S. (1995) *Images of the Educational Traveller in Early Modern England*, New York: E.J. Brill.

Withey, Lynn (1997) *Grand Tours and Cook's Tours: A history of leisure travel 1750–1915*, New York: Morrow.

Youngs, Tim (1994) *Travellers in Africa: British travelogues 1850–1900*, Manchester: Manchester University Press.

2   Limited Visions of Africa

Geographies of savagery and civility in early eighteenth-century narratives

Roxann Wheeler

The Negroe Town of Cape *Coast* is very large and populous. The Inhabitants, tho' Pagans, are a very civiliz'd Sort of People, for which they are beholding to their frequent Conversation with the *Europeans*. They are of a warlike Disposition, tho' in Time of Peace, their chief Employment is fishing, at which they are very dexterous.

(Smith 1744, 123)

He [Agaja, the king of Dahomey] was middle-sized, and full bodied; and, as near as I could judge, about forty-five years old: His Face was pitted with the Small Pox; nevertheless, there was something in his Countenance very taking, and withal majestick. Upon the whole, I found him the most extraordinary Man of his Colour, that I had ever conversed with, having seen nothing in him that appeared barbarous, except the sacrificing of his Enemies; which the *Portuguese* Gentleman told me, he believed was done out of Policy; neither did he eat human Flesh himself.

(Snelgrave 1734, 75)

Those People [Africans] in general are the most unpolish'd of the three ancient Parts of the World. Along the Coasts of the Mediterranean, where the Arabs formerly extended their Conquests, they are most civiliz'd, that Nation, renown'd in those Days, having still retain'd something of their former Government and more human way of living. The inner Regions, less known to us, as scarce ever frequented by other Nations, continue in greater Ignorance, and entire privation of all politeness; and the most Southern are altogether brutal, or savage.

(*The Compleat Geographer* 1709, 166)

Until Mungo Park's 1795 expedition into the African interior, there is little 'true' storyline about Africa available to Britons.[1] Instead, there are episodes of coastal contact in the early narratives, first for gold, ivory and decorative woods and later for slaves; many of these episodes yield confusion about the land and the inhabitants rather than clarification. Below, I will draw on several African travel narratives of the 1720s and 1730s, especially William Snelgrave's *A New Account of Some Parts of Guinea and the Slave-Trade* (1734).[2] I am particularly interested in juxtaposing an examination of these eyewitness accounts with a study of Daniel Defoe's *Captain Singleton* (1720), an early novel which depicts twenty-seven European men living on Madagascar and then journeying from Mozambique, across central Africa, to Angola with sixty African men whom they enslave to carry their baggage. Together these texts indicate the ideological limits of English writing about Africa and Africans during the first three decades of the eighteenth century. In these accounts, concepts of the 'European' and 'African' are still in formation and not solidified by racist ideology more typical of nineteenth-century colonial narratives.[3] Revealing the contradictions generated by previous frames of reference Englishmen brought with them versus the contemporary reality of Africa and trade, this set of narratives thus permits an analysis of the modes of othering and the simultaneous fracture of them through the pressure of trading situations.

As the epigraphs dramatize about English perceptions of Africa at this time, the investigation of place is inextricable from the reports about its people and from preconceptions about what makes one set of people different from another. The ordinary range of knowledge about Africa and Africans in the early eighteenth century was a combination of doubt and conviction, fact and fantasy. Although geographies were common in the libraries of the educated, it was through travel writing that most Britons gained their ideas about Africans and Africa. Indeed, in England, travel literature was second only to theological texts in popularity during the eighteenth century, and, although Africa was one of the least-known places, it figured prominently in the many compendiums of travel published during the eighteenth century.[4] In fact, almost four times as many books about Africa appeared in the first half of the eighteenth century as in all of the previous century (Hallett 1965, 137). Revisionist scholarship, such as Anthony Barker's history of the African image, has convincingly argued about this literature that 'this period before about 1769 was a fruitful one, yielding the most influential descriptions of Negro society of any in the eighteenth century' (1978, 17).[5] Specifically, it was the diverse regions of coastal western Africa that were the object of interest because of their strategic location for trade with Europeans, which was the primary reason for contact and the context for European narratives.[6]

One of the primary factors shaping English experience of Africa and the resulting narratives was the limited access Europeans had to Africa. Where the British and other Europeans had made inroads onto African land it was less a transformation than a small incursion: forts and factories stood on parts of the coast and were permitted by local leaders to facilitate the (slave) trade. Europeans paid rent and other tributes mutually agreed on, though if there was a change in power or in power relations between British and local middlemen, demands could increase considerably (Curtin 1964, 7–8). The Africans and Mulattos with whom the English traded were middlemen or local potentates – and the power dynamic was less polarized than in the colonies. The accounts of west Africa emphasize that European men were tenuously placed there and that they were engaged in commercial and political relations with local power bases rather than in anything approximating plantation management.[7] For instance, much of the trading, especially on the windward coast, took place on European ships or in temporary huts on the shore (Atkins 1735, 69). It is clear from these accounts that many Africans preferred not to allow Europeans on shore because, as most of these writers rightly believed, Africans wished to safeguard inland slaving and trading routes and to conceal the location of gold mines (Smith 1744, 138; Houstoun 1725, 24–5). In cases when they were permitted on shore, Europeans were carefully contained, and their experiences of the land and people were usually highly ritualized by the local powers.

On the continent, the British physical presence was small; very few Englishmen had contact with Africans in Africa: 'about 350,000 Britons had some measure of direct contact with blacks on the African littoral from 1600 to 1780' (Morgan 1991, 160). In any given year, there were about 15 English trading settlements with approximately 250 men (Davies 1957, 240). Coupled with the small number and vulnerable position of Europeans was a climate that held dangers for them at every turn. Until the advent of modern medicine, including quinine, Europeans faced widespread death in Africa. In fact, the west coast was grimly known as the white man's grave (Hallett 1964, 5) and accounted for the mortality of the newly arrived at the rate of between 25 and 75 per cent in the first year alone (Curtin 1964, 71). Despite occasional rumours of cannibals, the ferocity of the creatures that inhabited the rivers and nearby forests (men being devoured by crocodiles and sharks was a favourite story (Atkins 1735, 46)), not to mention the physical discomfort caused by the astonishing number of ants, flies and mosquitoes, and the constant misunderstandings between Britons and Africans which often resulted in violence, it was death from disease and fevers that kept European presence minimal in Africa (Curtin 1964, 71–7). In sum, coastal western Africa was not a place where Europeans could assert their authority easily.[8]

If Europeans could have made good on their power, they would have extended their presence into the interior. Significantly, the most frequently named frustration to British desire was the inability to know the African hinterland and to map it.[9] Until well into the nineteenth century, almost every book of world geography, travel narrative, or discussion about Africa mentions how little Europeans knew about it. As late as 1790, the editor of *The Proceedings of the Association for Promoting the Discovery of the Interior Parts of Africa* notes that 'the Map of its Interior is still but a wide extended blank' (1790, 6).[10] In contrast, contemporary maps commonly showed highly detailed coastlines, especially in the west, near the many European trading forts; this region was commonly spoken of in a way that reflected European and African interaction: trade on the grain, pepper, gold, ivory and slave coasts.[11] At that time, African geography was important to Europeans because of the land's produce and was renamed accordingly. To Europeans, western Africa literally signified the commodities integral to the profitable coastal trade. The interior, however, was usually sketched in, but with vaguer information. Sometimes, there were human or animal figures in the interior regions interspersed with mountains, lakes, or deserts. A map appended to Snelgrave's *A New Account of Some Parts of Guinea and the Slave-Trade* (1734) is representative of the way that fact and fiction mingled freely in information about Africa, even from men who had spent decades travelling there: rivers, tributaries, sizable lakes, main towns, mountains and kingdoms are delineated near the trading coasts and along portions of the Senegal and Gambia rivers. As the eye sweeps inland, one is apt to see, instead of terrain or location markers, discursive designations such as the Kingdom of Gago, 'From whence they carry Gold and Morocco'; the Kingdom of Bito, 'Whose Inhabitants are very Rich'; and the Kingdom of Temian, 'These People are Antropophagos [*sic*] or Men eaters'. As Snelgrave's map dramatizes, Africa had not fully taken shape for Europeans, and thus these early narratives are characterized by as keen an interest in African people as in the landscape; both interests, however, situate Africa as a place of trading activities, wherein potential is continually calculated.[12]

The lack of real knowledge did not inhibit a constant flow of rumour and speculation about the interior. Indeed, most writers complain of the contradictory information they received from both natives and other Europeans. Nevertheless, Europeans made frequent attempts to penetrate river routes farther inland, and, by the early eighteenth century, a few Englishmen had travelled several hundred miles up the Gambia river.[13] Two conflicting ideas about the interior were common: it was uninhabitable, barren and suitable only for the wild beasts that roamed there (Barbot 1699); it was a rich mine of gold, ivory and other natural resources which played a significant role in a newly consumer-oriented English society (Snelgrave 1734). Whichever construction was used, many writers believed that cannibals inhabited the interior.[14] Part of this tendency may be explained by the popular belief in England that frequent trading contact with Europeans rendered Africans more civilized.[15] Thus, the farther away people were from the coast, the more likely they were to be savage (see Atkins 1735, 123). Offering another perspective by arguing that 'the topic of land is dissimulated by the topic of savagery', Peter Hulme's *Colonial Encounters* (1986, 3) demonstrates the connections between reports of savagery and of gold. Resistance to Europeans combined with reputed wealth unleashed accusations of cannibalism; such a powerful mode of othering, developed in the Caribbean, was, at times, superimposed on Africa for the same reason. Since much was rumoured and little enough confirmed, the hinterland provided free rein to desire and fear.

Obstacles to penetrating the interior included the reputation of the inhabitants as much as the inability to get guides and access to those regions. All of the early eighteenth-century writers who actually went to Africa mention the extreme jealousy with which the Africans guarded information about the interior and access to it. Despite his own 'eyewitness' account of cannibalism, Snelgrave himself unwittingly reveals that the reputation of African barbarity and cannibalism enables African domination of the terms and conditions of trade, which included barring European access to the inland regions. For instance, he observes that on the windward coast, natives determine whether the trade will take place by smoke signals. Snelgrave repeats that Europeans are cautious of going on shore there anyway because the natives are allegedly 'barbarous and uncivilized'. Snelgrave mentions several times in revealing juxtapositions that he could never obtain satisfactory information about the interior of Africa and that fear largely inhibited the process: 'In those few places where I have been on shore my self, I could never obtain a satisfactory account from the Natives of the Inland Parts. Nor did I ever meet with a white Man that had been or durst venture himself, up in the Country; and believe, if any had attempted it, the Natives would have destroyed them, out of a Jealousy that they designed to make discoveries to their prejudice.'[16] Nowhere is the link between place and 'race' more evident.

Given the tentative relationship of Europeans to Africa itself and the primacy of trade relations, certain singularities arose when Englishmen regarded Africa. For example, historians have argued that the early eighteenth-century image of Africa and Africans was heavily influenced by fifteenth- and sixteenth-century contact with the first European traders and colonizers, the Portuguese, and by their accounts, which were still being reprinted (Barker 1978, 4). These accounts tended to focus less on the nature of the people and land than on (potential) trade relations. Before the Americas and the Caribbean had been settled, Europeans 'had seen African potential largely in terms of trade. Profits were made not by seizing the mineral resources of Africa but by invigorating existing African commercial systems' (Ibid.). The early eighteenth-century narratives feature both this older understanding and a newer one that assesses possibilities for improving, if not commandeering, Africa's natural resources.

Until the advent of the scientific explorer in the last quarter of the eighteenth century, whose stated purpose was disinterested knowledge not conquest, the travel narrators are connected to African trade, and the concern is to provide as much information as possible to enable it.[17] I am particularly interested in the way that trading concerns infiltrate the depiction of Africa. Historically, appreciation of nature or uncultivated land was not part of an early eighteenth-century sensibility; instead, signs of industrious human intervention constituted a notion of natural beauty. Thus it is little surprise that, although many writers comment on the incredible beauty of Africa, they are more interested in the plenty that the land produces. Atkins's appreciation is typical: at the Whydah fort, 'there are in the Bays of this River, variety of good Fish, that supplies the Scarcity of Flesh; Turtle, Mullet, Skate, Ten-pounders, Old-wives, Cavalloes, Barricudoes, Sucking-Fish, Oysters, Cat-Fish, Bream, and Numb-Fish; the most of which we catch'd in great numbers with our Searn; two or three Hours in a Morning supplying a Belly-full to the whole Ship's Company' (1735, 46–7). The attraction of the land is directly related to its abundance and its use to Europeans. Extensive lists of the available commodities or natural resources are common to early eighteenth-century narratives, a tendency of imperialist literature which Laura Brown identifies as the rhetoric of acquisition: 'the mere act of proliferative listing . . . and the sense of an incalculable quantity express the period's fascination with imperialist acquisition' (1993, 43).

Also typical of this time period is seeing the beauty of Africa as directly related to or the result of trade. For instance, James Houstoun sandwiches his frank admiration for the African countryside between a discourse on developing an indigo and soap manufacture and one on the best places to buy slaves. Specific places in Africa are important to the extent that they are sites of production to benefit potential English consumers.[18] Such a tendency helps to establish Africa as a place without history and thus without change in the European imaginary.

The most revealing English observations about Africa are those suggesting that it would be even more beautiful and lucrative, and the inhabitants even more civilized, through the active intervention, even insertion, of Europeans into Africa. The Dutch factor William Bosman, perhaps the most widely acknowledged European expert on Africa in the early eighteenth century, focuses on the labouring and ore potential of the mainland interior that is yet unexplored: 'For the *Negroes* only ignorantly dig at random, without the least Knowledge of the veins of the Mine. And I doubt not but if this country belonged to the *Europeans*, they would soon find it to produce much richer Treasures than the *Negroes* obtain from it' (1705, 86).[19] There are three significant concepts of difference operative in this passage that characterize most other accounts: Africans are technologically inferior to Europeans; their labour is squandered and incapable of fulfilling the demands of European markets; and Europeans would be better custodians of the natural resources.[20] Mary Louise Pratt remarks on a similar characteristic in John Barrow's *Travels into the Interior of South Africa in the Years 1797 and 1798* (1801): 'It is the task of the advance scouts for capitalist "improvement" to encode what they encounter as "unimproved" and . . . available for improvement' (1992, 61). She further argues that '[t]he European improving eye produces subsistence habitats as "empty" landscapes, meaningful only in terms of a capitalist future and of their potential for producing a marketable surplus' (Ibid.). This sentiment is a ubiquitous aspect of earlier narratives as well.[21]

Another significant factor shaping narratives about Africa was only partially about actual experiences and ways of seeing the land – it was the sense of cultural and visual difference evidently keenly felt by both Africans and Europeans. While the Bible and classical literature provided a long and influential frame of reference for perceiving Africa(ns), the origin of much of the primary information that comprises ideas about other societies came from male travellers to the East and Levant and from merchants, crews and physicians connected with the slave trade.[22] By and large, these men were empirically oriented, and many were dedicated patriots who had invested in Britain's economic strength and in the security of the empire. Their observations were avidly read and reappeared wholesale in compendious editions of travel narratives throughout the century as well as in magazines. Theologians, natural philosophers and other men of letters also included and refined upon these observations to illustrate their particular speculations. Thus, there was a widespread dissemination of this material about the characteristics of non-European people and of the British; it became part of everyday ways of thinking as much as it underpinned philosophical and scientific systems.

Connected to an increasing general interest in the rest of the world was the question of the origin of human differences; this question assumed a new urgency as Great Britain extended its empire. The dominant theory of the time, monogenesis, argued for the shared origin of humans descended from Adam and Eve and hence insignificant divergences based on geography or climatic differences. A very small minority of Britons referred to polygenesis to explain human differences; this theory also relied on the Bible, especially the Flood or the Tower of Babel, to account for the significant value of different colours and people because of their separate (i.e. inferior/superior) origins. A more secular theory about the cause of cultural differences established superiority based on the organization of familial, political and commercial structures, the use of technology, as well as the status of arts and sciences.[23] In formation during the first sixty years of the century, this model offered a progressive hierarchy from a primitive, nomadic hunting and gathering activity of isolated families to complex, commercial civilization. Its most basic assumption was that societies undergo development through successive stages based on different modes of subsistence and on the protection of private property (Meek 1976, 6).[24] Particularly useful for explaining the former barbarity of an advanced civilization like Britain, what came to be known as the four-stages theory also accounted for the elements of barbarism and savagery in Africa, America and Asia. In all aspects of economic and cultural production, differences of race and colour were thus increasingly central concepts with explanatory power.

Common ways of considering the cause and effect of differences were based on two major approaches: an older and more entrenched theory that understood racial characteristics to be fluid, formed over time by external forces working on the body (Schiebinger 1990, 393), and a newer theory that perceived racial char-acteristics to be more permanent, perhaps even innate, and which represented the mental or moral life of non-Europeans as significantly different from that of Europeans (Popkin 1973, 247). External theories unquestionably dominated during the eighteenth century. These external theories identified climate or geo-graphy as determinant, hence foregrounding accidental colour and behaviour differences as well as social customs which diverged from Britons' own. The epi-graph from *The Compleat Geographer* indicates the way that proximity to Europe and northern climates generated a theoretical hierarchy that was superimposed on Africa; because of the excessive heat which was believed to enervate the mind, morals and the body, commonplaces about the torrid zone being the home of indolence, lasciviousness and tyranny often seemed confirmed when Englishmen confronted polygamy and an organization of village life and labour that was alien to them. Since there was a great deal of contact that belied such sweeping general-izations, individual accounts often contain claims both about African barbarity and about polite trading exchanges that could have easily taken place between Europeans. In his *Travels into the Inland Parts of Africa* (1738), Moore, for instance, reports that 'The Natives, really, are not so disagreeable in their Behaviour as we are apt to imagine' (120). His statement attests both to the influence of received epistemology and to the tentative emergence of new know-ledge. Inasmuch as it was difficult for travellers to garner reliable information about Africa while they were there, it was equally difficult for Europeans to obtain consistent reports from travellers about the inhabitants.

Like their behaviour, the complexion of a people was believed to result from the heat of the sun; however, European exploration of similar latitudes elsewhere,

where they did not find inhabitants as dark skinned as the Africans, led many to seek alternative explanations. Although skin colour was not yet a primary constituent of racial ideology, European scholars and travellers often speculated on the singularity of Africans' complexion; no other people elicited such inquiry for the next two hundred and fifty years. Attempts to account for the origins and causes of different skin colours eluded Enlightenment scholars, but blackness remained an especially puzzling phenomenon. One of the early modern British inquiries into human complexion was Sir Thomas Browne's 'Of the Blacknesse of Negroes'. Browne's three chapters on colour in *Pseudodoxia Epidemica* (1646) comprise an extended meditation on the multiple causes of colour and variations even within a given country; they also rule out some popular misconceptions about the cause of colour. By presenting exceptions, Browne eliminates widely accepted single factors such as the heat of the sun, latitude or God's curse. Rejecting the dominant rationale of climatic determinism as an explanation for the variety of human colours, Browne speculates that the air and the proximity to rivers are also influential but are ultimately insufficient explanations. Browne finally concedes that 'However therefore this complexion was first acquired, it is evidently maintained by generation, and by the tincture of the skin as a spermaticall part traduced from father unto son' (56). In a related point, Browne consults conflicting ideas about beauty in mid-seventeenth-century Britain: some believe beauty 'to consist in a comely commensurability of the whole unto the parts, and the parts betweene themselves' (520–1). Others, however, 'and those most in number, . . . place it [beauty] not onely in proportion of parts, but also in grace of colour' (521).[25] Thus, at the time of major colonial expansion, complexion is just emerging as a culturally significant factor in the realm of aesthetics in England.

Within early scientific discourse, colour is one of a myriad of signs, but not more revealing than others. The relative importance of colour as a feature to observe is reflected in *The Philosophical Transactions of the Royal Society*'s directions to travellers about observing inhabitants of other countries: travellers were to record their general stature, shape, colour, features, strength, inclinations 'and Customs that seem not due to Education' (1667, 188). As the examples from Browne and the Royal Society suggest, discourses of aesthetics that identified beauty with whiteness were not commensurate with discourses of moral value or with practices of scientific observation in the late seventeenth and early eighteenth centuries.[26] Nonetheless, most writers concurred that visually, if not culturally, Africans (especially from the west, south and east) were distinctly different from Europeans.

Writing in England and producing one of the earliest racial taxonomies in 1721, Richard Bradley continued the precedent set by his seventeenth-century predecessors: colour divided populations, but it did not necessarily carry significant connotative weight by itself. Negritude was not self-explanatory.[27] Bradley was unwilling to associate physiological distinctions with mental differences: his fifth and last sort of man is

> Blacks of *Guiney*, whose *Hair* is *curl'd*, like the *Wool* of a *Sheep*, which Difference is enough to shew us their Distinctions; for, as to their Knowledge, I

suppose there would not be any great Difference, if it was possible they could all be born of the same Parents, and have the same Education, they would vary no more in Understanding than Children of the same House.

(Bradley 1721, 169)

Africans' colour has not yet emerged as the primary signifier of their identity or status.[28] In fact, one of the most fascinating indications of the relative fluidity of categories of difference at this time occurs when a black man claims that he is white because he is a Christian. In *Journal of a Voyage up the Gambia* (1723), the narrator describes their translator 'as Black as Coal; tho' here, thro' Custom, (being Christians) they account themselves White Men' (243). In detailing the group put together for the expedition, the narrator bows to custom and tells us that there were nineteen white men, including the black 'Linguister', twenty-nine black male servants and three female slaves to cook.

Although early racial taxonomies such as Bradley's highlight physical features and divide different populations by geography and skin colour, this was not yet a culturally dominant way of singling out others as inferior. Instead, Christianity still dominated English concepts of self versus Other. The flip side of the coin is that ideas of savagery dominated English concepts of Africans: nakedness was an especially important visual constituent of this difference. Nudity or, most usually, partial covering signified at once a less civilized and more impoverished society, a non-Christian (Adam and Eve had learned to hide their 'shame' by clothing) and, occasionally, a freer sexuality.[29] My point is that contemplation of differences occurred in a large matrix of inherited notions. In the 1720s, African slavery did not yet produce or require ideologies of race commensurate to Britain's involvement and profits at this historical moment, though their nascent traces appear in the travel narratives.

Given this argument about the cultural currency of certain differences, I wish to turn briefly to several trading scenarios in order to examine the tension between convictions of cultural difference and the dynamics of trade. Snelgrave's text is especially interesting since he had traded in Africa for decades, was staunchly committed to the slave trade and freely circulated rumours of African cannibalism; clearly he does not appear as a good candidate for a sympathetic depiction of Africans. Part one of his narrative consists of the political history of the rise of Dahomey power in the region of Whydah and its effect on trade.[30] Snelgrave minutely reports his conversations with the victor and the vanquished. Notwithstanding that these conversations take place through a translator and the reportage is based on Snelgrave's memory (i.e. it is highly mediated discourse), an intriguing phenomenon is still discernible. The extensive amount of conversation conveys not a uniform sense of English superiority, but a sense of the mutual jockeying for position and, in some cases, for understanding. Thus, as the epigraph from Snelgrave's text reveals, there is an odd mixture of mutual suspicion, cultural arrogance and occasional admiration.

In theory, each confrontation between Snelgrave and African potentates is an occasion to show the superiority of Christian morals, the barbarity of natives and the triumph of trade as the resolution to all difficulties. In the following example, a social visit is soon transformed into an opportunity for trade, in which Snelgrave brings into play the most typical sense of difference between the English and Africans. On an invitation from a chief man, Snelgrave goes to his village, where he sees a boy tied to a stake. Snelgrave inquires why he must die, and the king replies the boy must be sacrificed to his god Egbo for his prosperity. At this point a breach in hospitality occurs when Snelgrave orders one of his own armed men to release the child. In response, the king's man advances, also armed. Snelgrave removes a pistol from his pocket; at this point, the king rises in resentment and there is a heated exchange of words through an interpreter. The king accuses Snelgrave of violating the laws of hospitality, and Snelgrave finally concedes the breach: "'on account of my religion, which, tho' it does not allow forcibly taking away what belongs to another, yet expressly forbids so horrid a Thing, as the putting a poor innocent Child to death.'" Simultaneously confirming the cross-cultural sanctity of private property, which includes slaves, and notions of fair exchange (money for people), Snelgrave promotes an oxymoron: the benevolent slave trader. The king finally accepts payment for the child's life, and Snelgrave reassures us that he sells the boy to a good master in the West Indies. A successful transaction solves not only the tension of religious difference but also the trespass on hospitality. This fascinating encounter legitimates English participation in the slave trade through the intervention of Christianity: Christians' greater humanity preserves the life that Africans would sacrifice. Contradictorily, it casts African religion dismissively as pandering to a desire for gain that Christianity evidently does not foster.[31]

Other episodes, especially those concerned directly with trade, are more ambiguous, and the English assurance of superior difference is often deflated because, in reality, the English occupy a less empowered position than the Africans in trade negotiations. One of the most interesting scenes in Snelgrave's text is the settlement of future trading rights with the king of Dahomey. Snelgrave records the signs of a civil society as he travels inland: well-paved roads, populous villages and pleasant, cultivated countryside leading to the capital. There, the king is richly dressed in his court, and he receives gifts from Snelgrave. Clearly this is neither a naked savage nor the leader of what elsewhere Snelgrave refers to as a 'barbarous, brutish' nation. The king sits cross-legged on the ground, and Snelgrave must, out of courtesy and protocol, do the same, even though he finds it difficult (60). The negotiating that ensues appears to give Snelgrave the upper hand, but it is suitably uncertain, not least because it depends on Snelgrave being feminized in relation to the king – an almost unprecedented phenomenon in early English travel literature. Apparently, the king wants to encourage European trade now that the wars are over to help replenish his finances and secure hegemony in the region; he is a powerful conqueror and initiates the bargaining by expecting what the king of Whidaw formerly received. Snelgrave counters with flattery: since the king of Dahomey is so much greater than the vanquished king, he won't require as much from the Europeans. The king finally says to Snelgrave that 'he would treat me as a young Wife or Bride, who must be denied nothing at first' (62). Snelgrave is disconcerted by this wording but is led by the interpreter and the king to be encouraged, so he offers him half of what Whidaw received, and the trading 'marriage' is cemented. Finding the land, leader, army and inhabitants well organized, prosperous and brave, the sense of difference so obvious in the previous example is diminished by the myriad signs of power and civilization.

Even when the text highlights African difference, Snelgrave's narrative often does not maintain it. In the accounts, meetings with chief men typically elicit an uneasy combination of similarity and difference. As part of cultivating ongoing trade relations, Snelgrave invites one of the great men from Dahomey's court to visit him at the factory. Snelgrave serves him ham and minced meat pie on plates and offers silverware to use. Observing how awkwardly the African uses the fork, he finds it amusing. Clearly readers are meant to share this evidence of cultural superiority and find humour in the inferior manners of the African. Next, the great man asks how the pie is made and admires it immensely; Snelgrave replies that his wife made it and put it in earth pans to keep six months, even in this heat. The man then asks how many wives Snelgrave has, who says 'one'. The African laughs and owns he has five hundred wives and wished fifty could prepare the meat as Snelgrave's had done (79). On the one hand, this encounter is remarkable for what is rarely shown – a shared taste and a joke. The joke reveals a common ground for English and African men – a shared belief in women's subordination to men – but the joke is more at the expense of African women whose domestic skills do not compare to those of their English counterpart. On the other hand, undercutting this sense of similarity between the men is the polygamous Other who manipulates the protocols of politeness and *dashees*, or bribes, customary to maintaining trading privileges. Finally, this man admires the plates and silverware as well as the food, and Snelgrave feels that he must offer the single plate setting – and the chief takes them all, much to Snelgrave's irritation.

Snelgrave concludes the political history of Dahomey's rise with the happy ending of successful trade relations established in the region. He ends up with six hundred slaves and gets a good market in Antigua, where he loads sugar and returns to London. Yet Snelgrave mentions a litany of complaints about the ill usage he receives from the chief man of Jaqueen, who refuses to pay a balance owed to Snelgrave, and about his storehouses being plundered; this situation gives rise to accusations of their barbarity as a people. As all of these examples show, successful trade relations turn 'brutish' people into esteemed trading partners; a sour deal can reconfirm ideas about African barbarity. There is thus an intimate relationship between the significance of difference and successful trade, and this symbiosis carries over to Defoe's *Captain Singleton*.

These early narratives of voyages to Africa reveal it as a place of power politics between European and African men and among Africans themselves, as a coast controlled by African middlemen, and as a site of ritualized business dealings and trade constantly renegotiated because of European breaches of protocol and local power shifts. At this time, the constraints of the African trade frequently curtail contemporary systems of othering, but they do not revise them significantly; characteristically, they coexist. In turning to *Captain Singleton*, I wish to examine the differences as well as similarities when real contact and profits are not at stake. While on the one hand, Africans and trade are transformed quite dramatically from the available reportage, on the other hand, treatment of the land is similar to the desire present in other narratives, leading one scholar to speculate whether *Captain Singleton* was 'deliberately designed to stimulate interest in the potential value of the interior of Africa?' (Scrimgeour 1963–4, 23).[32]

In 1720, no European was known to have travelled more than a few hundred miles inland and certainly not across Africa. *Captain Singleton* has long been interesting to critics because of the overland African journey in the first half of the novel, and that portion of the novel has garnered the preponderance of attention over the years.[33] Not surprisingly, several scholars have insisted on the travel book quality of the novel (e.g. Watson 1952), and many have devoted their essays to source studies and to ascertaining what kind of geographical knowledge about Africa was available to Defoe.[34] Since Secord's study (1924) of Defoe's narrative method seventy years ago, most have concurred with his conclusion that Defoe shows an astute use of commonly available speculation about the interior regions.[35] In what follows, I am not as interested in the accuracy of Defoe's depiction *per se* as in the function of the overlaps and differences from contemporary accounts and in developing the arguments laid out in previous sections about the English cultural imaginary's confrontation with Africa(ns).

Although Singleton journeys across the continent and real-life contemporaries sailed up and down the coast with brief excursions inland, they all envision the surface of Africa as a place of obstacles through which they wish to move unimpeded. Although Singleton does not envision any greater transformation of the land than his living counterparts, the novel does alter the conditions of contact. European access to the coast and to the interior is largely unhampered by local challenges. There are no unsuccessful or tough trading negotiations and there is no question about who is in charge. European authority is easily asserted over the land, climate and people and regularly achieved through the use of guns and through technological feats such as building ships, keeping meat from rotting, removing salt from sea water to provide drinkable water and by coercing natives to work either as slaves or for a pittance. Neither are there any extensive trading towns, no mines and no Africans in charge of the routes over which they travel. Apparently not in contact with their neighbours, the inhabitants all live in primitive, isolated bunches with little discernible government. Defoe's novel envisions an Africa emptied of a commercial infrastructure, including trading and communication networks. People are no longer serious obstacles either: in the novel, the challenge is European survival in a hostile continent.

The bulk of the African interlude takes place in the interior, far from the main African–European trading centres; consequently, there is more scope for departure from contemporary reports of the west coasts (Scrimgeour 1963–4, 21). Given free access to the interior, what is discovered? Other than some generic changes in the terrain (fertile river beds, desert, vast lakes, mountains), punctuated by running herds of elephants, fierce leopards, the *de rigeur* crocodile attack and regular trading encounters, there is ivory – miles and miles of it – and gold. Defoe thus fuses the two main perceptions of the interior: it is at once uninhabited, barren in parts (during a 1,000 mile stretch of desert, there are no inhabitants and they nearly starve (105)), and a treasure chest. Singleton and the Portuguese repeatedly discover gold and ivory that the natives either do not know about or have no use for, a scenario that significantly belies the extensive African trade in ore as well as tusks and the reputation of certain African tribes for their artisan skills in jewellery and mask making. Remarkably, wherever Singleton discovers ivory or gold, there are no inhabitants to contest his taking them. This aspect reflects contemporary observations about the unrealized potential of the African landscape and what it could mean for Europeans.

Whether it was hundreds of miles of untouched ivory tusks or rivers full of unmined gold ore as in the novel, or fertile land aching to be tilled and sowed, many British men expressed more than a passing interest in realizing the land's potential that the Africans had apparently neglected. Ironically, many writers who had significant and long-term dealings with Africans frequently noted the superior agricultural abilities of certain groups – and this was one reason why specific nations produced such prized slaves.

The novel not only satisfies contemporary curiosity about the interior and fulfils the wildest speculations about its riches, but also toys with permanent settlement. The fact was that scores of eighteenth-century traders claimed that England had no interest in colonizing Africa, but this oft-repeated disavowal sits oddly with the observations about the greater ability of the English to plumb Africa's agriculture and mines. Defoe, however, made no such separation between opportunities for trade and plantation in Africa: '"Neither Africa nor America are yet fully discover'd: there are yet infinite treasures of trade and plantation, to be search'd after, innumerable nations not convers'd with"' (Earle 1977, 261, quoting *A General History of Discoveries* (1725–7)). Defoe, always interested in the potential national profits from colonization, believed Africa to be a promising site: Defoe 'argued that by studying climate as Crusoe had done, Englishmen could plant more effectively and, for example, bring the price of coffee and spices down. He saw Africa as the most promising place for this endeavor' (Backsheider 1989, 512).[36]

It was not Guinea, but Madagascar, where Defoe's fiction identified the most potential for English settlement. During the late seventeenth and early eighteenth centuries, Madagascar especially elicited European speculation about establishing plantations – not in any coherent way but in several off-hand statements in travel accounts and geographies.[37] These desires tell us a great deal about what Europeans thought important about themselves in contrast to the native inhabitants. Richard Boothby was among the first Englishmen to discern the potential of Madagascar, which he calls 'the paradise of the world' (1646, 633). Boothby believes that the fertile island could be wealthy, 'especially if once inhabited with Christians or civil people, skillful in agriculture and manufactures, and all sorts of mechanic arts and labours' (633); 'It is a great pity so pleasant and plentiful a country should not be inhabited with civilized people, or rather Christians; and that so brave a nation, as to person and countenance, only black or tawny, should be so blindly led in their devotions' (635). Defoe arranges for the Europeans to settle there upwards of a year and coexist with the natives. Thus, it is little surprise when Singleton himself believes Madagascar the best place in the world to settle: 'I confess, I liked the Country wonderfully, and even then had strange Notions of coming again to live there; and I used to say to them [the Portuguese men] very often, that if I had but a Ship of 20 Guns, and a Sloop, and both well Manned, I would not desire a better Place in the World to make my self as rich as a King' (36). The novel proves the 'truth' of this maxim repeatedly: in several scenes European men obtain wealth through violence and through the transportation capability of maritime technology. Moreover, Madagascar had already proven its profitability to the British empire: of the 32,000 slaves on Barbados at the end of the seventeenth century, half originated from Madagascar (Kent 1992, 864). The island was renowned not just for its potential profit to European settlers or as an alternative source of slaves when the Royal African Company had inadequate supplies, but also as a pirate stronghold. Pirates 'had frequented African shores as early as 1691; by 1718, Madagascar served as both an entrepot for booty and a spot for temporary settlement' (Rediker 1987, 257). To the British, Madagascar in particular represents potential for settlement as well as expanded trade in Africa. Singleton's sense that violent intervention was the only way to settle successfully (i.e. profitably) reflects a desire, not so much to settle Africa with Europeans, as to control the flow of goods to Europe.

Defoe not only flirts with plans of settlement and greater profit-making but also imagines reinvigorating formerly profitable trading areas by showing the largest caches of ivory and gold only a few hundred miles inland from a coast already familiar to Europeans – and undervalued by them. Singleton and the Portuguese emerge from their overland journey in Angola with tons of ivory tusks and gold extracted and transported by the slaves. Angola, along with the Congo, was the only region where it was necessary for Europeans to journey inland to conduct business (Hallett, 1964, 7), and Defoe suggests that daring to go farther would lead to significant profits – with native cooperation. This region was predominantly a slave trading region; in fact, Davies claims that 'slaves were the only significant product' there (1957, 45). In contrast, at places farther north, gold, ivory and wood were also important items of trade. To the Europeans, this was the most remote commercial region, and it was a Portuguese sphere of influence which had declined in profitability since the mid-seventeenth century (Ibid., 231–2). Defoe transforms it into an area ripe for expansion and valuable for much more than just slaves.

If Defoe's novel rehearses certain cultural fantasies of expanding the empire through plundering Africa and describes a continent ripe for English commercial development, the novel underestimates African people proportionately. Instead of an agriculturally based society with extensive trading concerns and frequent warfare (like England), Defoe offers a more primitive society of scantily clad cowherds. Unlike travel narratives, *Captain Singleton* produces the greatest sense of difference regarding trade, notions of civility versus savagery, and labour.

In fact, Defoe critics have noted the difficulty of recovering much useful information in the novel about early eighteenth-century views of Africans. For instance, one critic complains about the inscrutability of the text and paucity of 'real' information about Africa and Africans in the novel: 'They are, in fact, almost identical with the South American cannibals of *Robinson Crusoe* and the islanders of *A New Voyage*, and there is nothing here but the most superficial and elementary of pictures' (Scrimgeour 1963–4, 26).[38] Laura Brown agrees with such an assessment: 'Because they are not seen as human, their difference, racial or otherwise, from the European adventurers is at first barely recognized' (1993, 162). Ironically, this superficiality produces apparent legibility: 'their characteristics are so repetitive that they are easily cataloged' (Scrimgeour 1963–4, 26). It seems to me, however, that the novel defeats such ease of decipherability precisely because of the contradictions noticeable in many trading encounters and in the use of 'savage', 'naked', and 'slave'.

In the novel, one of the main ways of indicating the state of civilization and conveying positive responses to other people is through their willingness to trade with Europeans (Brown 1993, 163–4). Essentially, there is a geography of civility: all travellers concurred that coastal people who traded with Europeans were more civilized than the inland tribes. According to this standard, the narrative represents natives of non-European countries as 'Barbarians' if they 'will not allow any Trade or Commerce with any *European* Nation' (238). Indigenous people, whether they refuse to trade with the Europeans or support their presence, meet with similar treatment – casualties by the score (213). The issue in both travel narratives and in *Captain Singleton* is native inhabitants who wish the Europeans to leave and the negative European response to thwarted trade.

Defoe's treatment of the different villages that Singleton and the Portuguese encounter in Madagascar and between Mozambique and Angola in the west may be gauged by their alacrity to trade and their ingenuity for it. On the one hand, the northern people of Madagascar are considered courteous (1720, 37, 39): Singleton's perception of Malagasy civility rests on their recognition that the Europeans 'were in Distress for want of Provisions' (37). In fact, civility in Madagascar seems equivalent not to trade but to cost-free supplies; these people give the Europeans goats and steers and refuse to accept anything in return – a situation unparalleled in any other contemporary account. Moreover, they offer other food as well as their labour freely to the Europeans: a captain of the natives 'seeing some of our Men making up their Huts, and that they did it but bunglingly, he becken'd to some of his Men to go and help us. . . . They were better Workmen than we were, for they run up three or four Hutts for us in a Moment, and much handsomer done than ours' (38). On the other hand, natives in southern Madagascar are perceived as civil only out of fear of the armed Europeans (14). They are worse than the natives from other places: 'scarce human, or capable of being made sociable on any Account whatsoever' because initially they do not engage in trade with the Europeans on European terms (21). Nevertheless, the narrative transforms these very same natives into 'very civil' people when later they bring food to the Europeans and try to barter (22, 27). In these examples, civility is determined through willing interaction and trade with Europeans, yet civility also connotes shrewd trading, even boldness, in a colonial context. For instance, in the novel, the Africans accustomed to European trade

> were a more fierce and politick People than those we had met with before; not so easily terrified with our Arms as those, and not so ignorant, as to give their Provisions and Corn for our little Toys. . . . But as they had frequently traded and conversed with the *Europeans* on the Coast, or with other Negro Nations that had traded and been concerned with them, they were the less ignorant, and the less fearful, and consequently nothing was to be had from them but by exchange for such things as they liked.
>
> (Defoe 1720, 122)

Europeans reaped less benefit from 'Europeanized' Africans, even if they admired them more for reflecting aspects of themselves that they prized most.

The novel also offers an alternative interpretation of the effect of European contact on Africans. Singleton observes that some Africans who have contact with Europeans have been cozened by them so that they are hostile and corrupt, not civilized: 'those who had seen and trafficked with the *Europeans*, such as *Dutch*, *English*, *Portuguese*, *Spaniards*, &c. that they had most of them been so ill used at some time or other, that they would certainly put all the Spight they could upon us in meer Revenge' (108). There are, in fact, three competing versions of African civility: contact with Europeans has rendered Africans more civil, but it may exist at the cost of European economic domination; Africans' civility signifies their ignorance of trading and technology – the most 'courteous' and 'good-natured' inhabitants are also the most ignorant of exchange value (122); and that European contact is pernicious, not beneficial to either Africans or Britons.[39] In fact, hunger for quick profit often led to one English trader's behaviour shutting down an entire region for trade. Taken together, these examples suggest that as much as Britons glorified trade, trade was not always beneficial to others in quite the way it was believed to be. At moments like this, critique of dominant ideology is available though the presence of contradictory statements.

As I've intimated, it is not only trade *per se* that defines one people as civil and others as barbarous: the trade must acknowledge a European concept of value. *Captain Singleton* and some other narratives about Africa (and later about the Pacific islands) emphasize the great gains Europeans can obtain for 'Trifles'; that is, natives do not seem to possess any notion of commensurate exchange value with Europeans. For example, Singleton and the Portuguese, stranded in Madagascar, are initially embarrassed when the natives bring them food:

> we were in the utmost Confusion on our Side; for we had nothing to buy with, or exchange for; and as to giving us things for nothing, they had no Notion of that again. As to our Money, it was meer Trash to them. They had

no Value for it. . . . Had we but some Toys and Trinckets, Brass Chains, Baubles, Glass Beads, or in a Word, the veriest Trifles . . . we might have bought Cattel and Provisions enough for an Army.

(Defoe 1720, 27)

The Cutler in their group spends a fortnight converting their money into bracelets and necklaces of metal birds and beasts.[40] Singleton patronizingly observes that they 'were surprized to see the Folly of the poor People. For a little Bit of silver cut out in the Shape of a Bird, we had two Cows' (28). A similar complaint about the Madagascar natives exchanging fresh beef for dried beef suggests their stupidity: 'they were so pleased with it, and it was such a Dainty to them, that at any time after they would Trade with us for it, not knowing, or so much as imagining, what it was; so that for Ten or Twelve Pound Weight of smoked dry'd Beef, they would give us a whole Bullock' (38–9). The higher the stakes, the more gleeful the narrative is about the natives' alleged ignorance: 'our Artificer found a Way to make other People find us in Gold without our own Labour'; he 'sold his Goods at a monstrous Rate; for he would get an Ounce of gold, sometimes two, for a Bit of Silver, perhaps of the Value of a Groat, nay if it were Iron; and if it was of Gold, they would not give the more for it; and it was incredible almost to think what a Quantity of Gold he got that Way' (136).[41]

Many twentieth-century critics have emphasized *Captain Singleton*'s almost exclusive focus on trade, especially on the acquisition of gold, and Laura Brown has pointed to this anachronism in the novel as obfuscating the trade in human beings (1993, 162). *Captain Singleton* is set in the 1690s; even then, gold had become incidental to the slave trade, except in Sierra Leone (Rodney 1970, 154). Historically, the shift in the 1640s to sugar plantations in Barbados (and later in the century elsewhere in the Caribbean) led to more attention to the slave trade and to the intensification of it. It was not until 1663, for example, that the slave trade was first mentioned as an objective of the Company of Royal Adventurers, the predecessor of the Royal African Company (Davies 1957, 41).[42] In all of the scenes where value is an issue, two aspects of trade with Africa are erased: the European labour that provides adornment or the novelty of dried beef and African cultural dynamics that value local status display. The natives trade cattle for silver bracelets of carved animals or dried beef because they are already fulfilling basic needs the Europeans are not: they trade foodstuffs for luxury items – local or imported ones with value-added European labour. Throughout the novel, the trading power dynamic is shown as much more polarized than it was, especially in relation to gold. Remarkably, the novel offers only a one-way flow of gold – from Africa to Europe; in fact, Europeans traded gold with local businessmen. Atkins, for one, reports that to appease a lapse in payment for watering English ships, John Conny accepts six ounces of gold and an anchor of brandy (1735, 76).

Defoe's picture of European trade with the Africans in the novel was not commensurate with European views of the Guinea trade, as I illustrated above. The narrative depicts situations in which Africans are made to appear more childlike and ignorant than they appear in contemporary accounts of trade. In fact,

Europeans frequently complained of native non-cooperation, hard bargaining and savvy business sense on the west coast.[43] Moreover, modern historians of the African trade have noted that 'For both Europeans and Africans, the numerous items of trumpery were placed at the bottom of the scale of values. It was said in 1607 [in a travel account included in *Purchas his Pilgrimes*] that bells, garlic, and other "trifles" and "toyes" were by themselves incapable of purchasing anything but foodstuff' (Rodney 1970, 172) – as depicted in *Captain Singleton*. On the other hand, 'Weapons stood at the other extreme from trinkets' (Rodney 1970, 173): Africans demanded them and some Europeans feared their passing into African hands. Private traders and pirates were mainly responsible for selling heavy weapons to African rulers in the late seventeenth century; at that time, a trade in swords was more acceptable to Europeans (Ibid., 174). It was the Dutch, English and French who introduced firearms in any quantity in the later seventeenth century.[44] By 1730, 'the annual imports of guns into West Africa had reached the figure of 180,000' (Wolf 1982, 210).

Such an apparent contradiction between the novel's representation and trade relations also appears in Defoe's non-fiction about the African trade. While he remarks the ignorance of Africans who trade bullion for trinkets, he also reveals the 'terms of trade' by which Europeans secured such profits. Because the Portuguese who first settled in Africa

> found the Natives, Wild, Barbarous, Treacherous, and perfectly untractable as to Commerce, and therefore to maintain the Trade, which they found profitable, they made little Settlements there. . . . They found it at last necessary to fortify themselves, and maintain their Possession by force . . . [this lesson has] taught us, *if we please to learn* this Maxim in the *African* Trade, that is no way to be carryed on but by Force; for a mere Correspondence with the Natives as Merchants, is as impracticable, as it would be if they were a Nation of Horses.
>
> (Defoe 1709, 560)

Because it is so intimately a part of trade, violence secures profits and underwrites British 'civility'. In fact, the British had the worst reputation among Europeans for deceiving and kidnapping locals (Atkins 1735, 151).

Despite the frequent trading encounters in *Captain Singleton*, Africans' savagery and nakedness are their most frequently cited qualities; Africans' savagery does not include cannibalism, as it does in Snelgrave's text, nor does it conjure polygamy or fetish worship as in contemporary narratives.[45] In the novel, 'savage' refers less to real fierceness or hostility toward Europeans than to those who simply are not European. For example, 'savages' applies to Africans who are known and friendly as well as unknown and threatening. Compare the following representations: 'The second Journey he went, some more of our Men desired to go with him, and they made a Troop of ten white Men, and ten Savages' (134); and 'At one of the Towns of these savage Nations we were very friendly received by their King' (130). As the first example dramatizes, there were at least two unequal categories of men: white ones and savage ones. That savagery had assumed a generic quality is reflected in the narrative's reference to the inhabitants of Africa and Australia as 'Indians', by which a savage is meant. Because 'Indian' became a synonym for savage, the term was used in the eighteenth century to refer to any group of native peoples considered primitive by European standards. Strangely, the Africans' European reputation never matches the characters' encounters with them; for example, upon first arriving on Madagascar, Singleton reports: 'At our first coming into the Island, we were terrified exceedingly with the Sight of the barbarous People; whose Figure was made more terrible to us than really it was, by the Report we had of them from the Seamen' (14). On the continent of Africa, Singleton initially notes that 'we had Nations of Savages to encounter with, barbarous and brutish to the last Degree' (48). While the novel cites and conjures savagery to signify Africans, it neither does so with as many resources as it might have nor with much textual evidence. In fact, Scrimgeour remarks dryly that Defoe 'does not exert himself to use what information was available' (1963–4, 29). Despite repeated narrative encounters that demonstrate the civility of Africans, conventions associated with trade have not yet triumphed over previous frames of reference in the novel.[46] Africans are 'civil' because they trade with the Europeans, yet they remain savages because of their poor sense of value and their underclothed bodies.

Lack of clothing, rather than skin colour or particular customs, signifies Africans' alleged savagery in the novel: 'These People were all stark naked' (Defoe 1720, 118) appears over and over again.[47] The effect of African nakedness is to emphasize European superiority, especially their technological superiority over brute physical power:

> we found them a fierce, barbarous, treacherous People, and who at first look'd upon us as Robbers, and gathered themselves in Numbers to attack us.
>   Our Men were terrified at them at first . . . and even our black Prince seemed in a great deal of Confusion: But I smiled at him, and shewing him some of our Guns I asked him, if he thought that which killed the spotted Cat . . . could not make a Thousand of those naked creatures die at one Blow?
>
> (Defoe 1720, 73)

As the narrative reveals, nakedness is not just a sign of cultural inferiority to Europeans, it has an important material aspect: nakedness reduces Europeans' potential gain. 'There was no great Spoil to be got, for they were all stark naked as they came into the World, Men and Women together; some of them having Feathers stuck in their Hair, and others a kind of Bracelets [*sic*] about their Neckes, but nothing else' (77).[48] As the example suggests, little was to be gained from conquest in which land was not the objective and where the people possess little in the way of weaponry or personal adornment of value to Europeans.[49] Rather than signifying a potential market for British woollens or re-exports of India cloth (as in fact was the case), their nakedness represents a lack. Such an absence in the depiction of Africans permits an even greater sense of their distance from Europeans, since they seem to possess no other distinguishing cultural traits, religion, or history – and it encourages a sense of African vulnerability.

Although references to savagery and nakedness are the central discursive indications of Africans' difference as the Europeans cross Africa and engage in trade, the focal point is not so much the different villages of 'strange' Africans encountered as much as the sixty men whom the European men enslave – especially the Black Prince. There are some significant narrative avoidances noticeable in the treatment of slaves, a characteristic which offers a different perspective on the status of trade in English society. Even though some Britons did not believe they should be a slave-trading nation, there was a prevailing sense of the greatness of the newly forming nation and empire; anything that added to its wealth and bested their European rivals was a boon. One of the narrative singularities is that neither 'slave' nor 'servant' is used in relation to the Africans who are forced to accompany the Europeans across the interior of Africa (see 54, 73 for exceptions). Avoiding terms that signified labour for others and subordination to Europeans is strangely reinforced by the narrative refusal to connect English agency with the taking of slaves – an interesting effect since elsewhere Defoe shows no such reluctance.[50] In the novel, the first instance of disassociating English agency from the slave trade translates English aggression immediately into native perfidy, which then justifies European armed defence – and the 'happy' result: taking slaves legitimately. After Singleton's designing to provoke a quarrel to take slaves 'legitimately' (i.e. as prisoners of war), a proposal rejected by the Portuguese, 'the Natives soon gave them Reason to approve it' (51) because they cheat in trade. There are two significant features of this scene: Singleton's initial design for acquiring slaves is not approved by the Portuguese because there is apparently no just cause. Secondly, the same desire is rapidly fulfilled by rechannelling it and cathecting it to a legitimate cause: native treachery, not European aggression. After initial resistance, the African men willingly, even though handcuffed, aid the European enterprise. This scene simultaneously avows and disavows the way most slaves were obtained – by deliberate warfare, which was sometimes fomented by Europeans – but Defoe has inserted Singleton and the Portuguese into the place of the Africans who procured slaves. Not least, Europeans are now in charge of procuring slaves inland and selling them at the coast; they have gained limited control over the supply of slaves. The novel presents both the original scenario of violent European desires for procuring African labour by any means available and its cover story: the methods of obtaining slaves are legitimate and their attitude towards labour is cheerful, not resistant.

Another example of altered discourses when slaves are the issue is the preferred term used to refer to the Africans whom Singleton and the Portuguese enslave: 'prisoners'. Appearing insistently, especially in the beginning of the journey across Africa, this term functions to legitimate the Europeans' enslaving Africans and suggests a temporary status rather than the permanent bondage customary in the colonies.[51] Singleton invokes the 'Law of Arms' to justify forcing the African men to labour:

> It presently came into my Head, that we might now by the Law of Arms take as many Prisoners as we would, and make them travel with us, and carry our Baggage. . . . Accordingly we secured about 60 lusty young Fellows, and let them know they must go with us; which they seemed very willing to do.
>
> (Defoe 1720, 54)

As John Richetti suggests, 'The "Law of Arms" is a euphemism made possible by the narrative's conversion of a nakedly aggressive situation into a purely defensive and necessary one' (1975, 84). But the Law of Arms applies only to Africans supervised by Europeans, not to African actions against European violations. By transforming the men into prisoners, the novel makes slavery seem temporary, even contractual; further, the novel links this incident to the most common way that Europeans obtained Africans for sale in Africa: 'prisoners of war made up a large, if not the greater, part of the supply of slaves sold to the Europeans' (Davies 1957, 227).[52] Although in the novel the sixty Africans prove immediately trustworthy and even cheerful, these slaves are initially considered less compliant than the people of Madagascar: 'fierce, revengful and treacherous, for which reason we were sure, that we should have no Service from them but that of meer Slaves, no Subjection that would continue any longer than the fear of us was upon them nor any Labour but by Violence' (54). Even disregarding the noticeable contradictions about these peoples' nature and willingness to work for Europeans (and on the same page as well), the novel justifies their violent subjection at the same time as it expresses a desire for willing servants rather than 'meer Slaves'.

In another case of disassociating European agency from the slave trade, Singleton and the other pirates literally run into a ship full of slaves with no European masters (155–6): the question is not whether to free them but which nation to sell them to (164). Enormous profits from illegal sales of slaves to the Portuguese in Brazil accrue in the same way profits from gold accrue in Africa: the land and sea simply offer up bounty for the taking. There is no little irony in this successful fictional transaction because Defoe proved absolutely committed in his *Review* to arguing on behalf of the Royal African Company as the most reliable provider of slaves to the English plantations. Despite Defoe's belief in the benefits of monopoly, interlopers or private traders, sometimes equated with pirates in the early eighteenth century, were the means by which Britain 'developed and eventually dominated the slave trade' (Barker 1978, 9). As all of these examples reveal, even though trading in slaves was a largely unquestioned practice at this time, narrative acrobatics to make exploitation appear more legitimate than convenient to the Europeans are noticeable; they demonstrate a certain cultural anxiety about equating humans with commodities.

In this vein, the Black Prince commands special notice because from his first introduction, the narrative positions him midway between the Europeans and other Africans. Because of his illustrious lineage and high status with his countrymen, Singleton makes use of him as an overseer and mediator:

> There was among the Prisoners one tall, well-shap'd handsom Fellow, to
> whom the rest seem'd to pay great Respect, and who, as we understood
> afterwards, was the Son of one of their kings. . . . As I found the Man had
> some Respect shew'd him, it presently occured [*sic*] to my Thoughts, that
> we might bring him to be useful to us, and perhaps make him a kind of
> Commander over them.
>
> (Defoe 1720, 57)

Singleton's willingness is based on the prospect of gain and 'evident signs of an
honourable just Principle in him' (62). Distinguishing the prince, however, follows a pattern typical, not of European brotherhood, but of plantation management: elevating one or more male slaves to a supervisory position. 'The person
with the most authority and greatest responsibility among slaves – at least in the
workplace – was the driver, foreman, or ranger. . . . Drivers received many favors
not shared by other slaves' (Morgan 1991, 201). It is this position as overseer,
rather than any kind of 'whitening' or European facial features, that marks the
prince's difference from his fellow Africans. Throughout the eighteenth century,
the British remain compelled by domestic notions of rank in a colonial situation.
Although it is tempting to interpret the prince's narrative position as a result of
European recognition of individual merit or of honouring local hierarchies of
rank, the novel emphasizes the expediency of exploiting local power relations to
ensure cooperation in gaining profit, as in fact was the case along the coast.[53]

The prince is both obedient and manly, even comparable to a Christian in his
behaviour:

> [N]ever was Christian more punctual to an Oath, than he was to his, for he was
> a sworn Servant to us for many a weary Month after. . . . I took a great deal
> of Pains to acquaint this Negroe what we intended to do, and what Use we
> intended to make of his Men; and particularly, to teach him the Meaning of
> what we said . . . and he was very willing and apt to learn any thing I taught
> him.
>
> (Defoe 1720, 59)

Unlike the men whom Snelgrave and Atkins encounter, the prince's masculinity is
not threatening because it is underwritten by his subordination to Singleton
especially: 'he made great Signs of Fidelity, and with his own Hands tied a Rope
about his Neck, offering me one End of it' (61). Despite this submission, the
prince also has a modicum of influence over the Europeans: 'He [the prince] took
some of the things up in his Hand to see the Weight, and shook his Head at them;
so I told our People they must resolve to divide their Things into small Parcels,
and make them portable' (59). The prince thus 'negotiates' some terms of labour.
This scene represents one of many in which the prince's position is intermediate
between Europeans and other Africans.

The prince's difference from the other Africans is also apparent in the
Europeans' different labour expectations for him: 'I made Signs again, to tell him,

that if he would make his Men carry them, we would not let him carry any thing' (60). The prince relays his men's needs to the Europeans and vice versa (60, 65): he is the conduit for maintaining stable race relations, i.e. willing subordination to the white men.

One of the most striking ways that the Europeans distinguish the Black Prince is that they consider him an agent worthy of being granted autonomy. At the end of the African journey, the prince's fate differs from that of many of the others: 'The *Negro Prince* we made perfectly free, clothed him out of our common Stock, and gave him a Pound and a half of Gold for himself, which he knew very well how to manage' (137). Literally, he is valued at a little less than half a European, all of whom receive four pounds of gold! (Previously, when the Europeans had been rewarded with three and a half pounds of gold, the prince received one pound (97).) European clothing, understanding the value of gold and freedom of movement define Europeans and the prince. The nameless others do not fare as well: some 'of our Negroes as we thought fit to keep with us' (137) were not released to return to their homeland; in fact, Singleton takes two Africans with him – whether to Cape Coast Castle or to England is not made clear (137).

Such a production of two different 'types' of Africans explains Europeans' split behaviour: the enslavement of a vast majority and the dealings with middlemen and local powers as almost equals. This second group was assimilated into European consciousness chiefly by means of domestic paradigms of station, masculinity and manners (civility). As the representation of the prince dramatizes, the novel relies on treating a majority of Africans as faceless masses who labour – not because they are pagans, black, or lacking in Western notions of art and letters, but because it benefits Europeans. European gain and not simply African lack commands attention in *Captain Singleton* and in other accounts. The ideology of race as we are familiar with it today does not excuse European aggression in the novel or indeed in any publications of the 1720s; aggression is a manly part of consolidating imperial power. At this time, advantageous trade was sufficient reason for engaging in the buying and selling of humans; there was also no systematic way to justify slavery because, as I've demonstrated, systems of othering were still in the process of developing. In other words, in the early eighteenth century, the preferred British justifications for enslaving Africans did not centre primarily on ideas about racial inferiority, rather on African overpopulation of their continent, their historic practice of enslaving prisoners of war and their usefulness to British sugar plantations especially.[54] None of this means, however, that European exploitation of Africa was not mediated and palliated through narratives such as *Captain Singleton* and its characterization of 'legitimate' slavery and beneficial commercial contact or of the Africans' 'failure' to realize their land's potential.

In analysing the ways in which the novel produces a visible economy of race for European and African men that does not rely primarily on differences of colour or bodily features, but on ideas about clothing, commercial behaviour and manual labour, I have emphasized the significance of violence (threatened or enacted) in securing European dominance and its visible presence in underwriting relations of trade and labour. I have also argued that the novel represents a picture of Africa that dovetails with contemporary British commercial desires. It is a land that provides the raw materials (gold, ivory and food) and labour power for European profits; its 'place' in an increasingly global economy is not that of a nation of consumers or of manufacturers, but that of a nation which does not value consumption. The repeated scenes of asymmetrical trading, Africans' lack of clothing and the negligible spoils for Europeans from conflict all align Africa on a dangerous side of the imperial dynamic. Thus, Africa and its peoples fit into the emerging British empire as resources for exploitation secured by superior technology and knowledge. Unlike travel narratives, *Captain Singleton* reveals, even as it disavows, the structures of power that secure and promote selective representations of European–African contact.

NOTES

1 Curtin (1964) observes about the way Africa signified in the eighteenth as opposed to the nineteenth century: 'very little was known about the interior of North or South America or about Central Asia [either]. The image of "darkest Africa", either as an expression of geographical ignorance, or as one of cultural arrogance, was a nineteenth-century invention' (9). Two insightful analyses of nineteenth-century exploration narratives and the emergence of the dark continent are Brantlinger's 'Victorians and Africans' (1985) and Duncan's 'Sites of representation' (1993), especially 47–54.

What is distinctive about the depiction of Africa in pre-abolition narratives is that it does not figure as a *locus amoenus*, a setting for exotic adventures, as an example of an ancient civilization and empire, or as a land peopled by noble savages. Parts of the Americas, the Orient, East Indies and, later in the century, Tahiti were more typical of these constructions.

2 The bulk of events that Snelgrave features occurred roughly between 1713 and 1726. Historically, these texts are contemporaneous with the reinvigoration of the Royal African Company in 1719, when the Duke of Chandos infused money into the lagging company, which was losing out financially to separate traders. The Royal African Company, chartered in 1672, was a monopoly organization to trade slaves and other goods with western Africa; because of the great profits, interlopers had been common almost from its inception. To illustrate the upsurge in trade, in 1712, 33 ships left England for Africa; in 1726, 200 were engaged in trade. By 1720, yearly exports to Africa averaged £130,000 (Hallett, 1964, 8). The 1720s is thus an especially important time to analyse narratives since there is a major increase in the profitability of the slave trade, and its importance to notions of British national prosperity increases accordingly.

After the conclusion of the War for Spanish Succession in 1713, the British gained the Asiento from Spain which gave them pre-eminence in supplying slaves to the Americas. At this time, imperialism rapidly expanded and there was an increasing orientation toward the trade or commercially based version that served the interests of a pre-industrial capitalist society. It was a mercantile system characterized by a national policy of protectionism; chief concerns included a favourable balance and terms of trade, coercive control over colonial production and commerce, accumulation of precious metals, support for domestic industry and hostility towards the other major imperial powers – Holland and later France.

3 Jablow also considers the issue of racialized categories applicable to Africans and Europeans (1963, 2ff). 'Africa' and 'African' were not transparent or monolithic terms in the eighteenth century; our late twentieth-century notions have no parallel in the seventeenth and eighteenth centuries. I use the term most frequently as a substitute for 'Black' or the eighteenth-century terms 'Negroe' and 'Native'. For ideas about the European invention of Africa versus heterogeneous definitions of Africa, see Appiah (1992, 25) and Mudimbe (1988, 69). 'Africa' and 'African' have no less heterogeneity today, but ideologically and geographically many Westerners have created a more unified picture of Africa than its various histories, customs, religions, and languages demand. Novels such as *Captain Singleton* are part of this homogenizing tradition.

   As opposed to 'African', current since the ninth century, 'European' was a relatively new term; the *OED* lists the first citation in the early seventeenth century. Its first meaning reveals the way its usage emerged in response to colonialism. 'In India, *European* (not "English" or "British") was the official designation, applied to the troops sent from the United Kingdom as distinguished from the native soldiers.' 'European' also signified a 'person of European extraction who lives outside Europe; hence, a white person, especially in a country with a predominantly non-white population'. The first entry occurs in 1696.

4 For an excellent analysis of the cultural desire to consume fantasy, knowledge and literature about Africa, see Nussbaum's 'The other woman: polygamy, *Pamela*, and the prerogative of empire' (1994, 138–46). In a more straightforward way, late seventeenth- and early eighteenth-century educational curricula increasingly included geography, which meant that there was often very detailed information about many other populations, including Africans, being taught. 'In the eighteenth century, while the classics continued to predominate, geography was increasingly recommended by educationists, most often as an essential ancillary to history. In some schools, notably Rugby, as well as figuring in the syllabus, geographical works were regarded as the most appropriate form of leisure reading, particularly for younger boys. In dissenting and private academies, where the curriculum was less oriented to the classics, geographical studies played an even larger role. At the same time contemporary advertisements, seeking and offering private tuition, showed that geography was a common ingredient of education outside the schools, not least for girls' (Barker 1978, 22–3).

5 The history of British commercial contact, imperialism and racism has resulted in several insightful studies of the literature that helped shape and resulted from these phenomena. The best survey of individual accounts is found in Marshall and Williams *The Great Map of Mankind* (1982). Excellent analyses of eighteenth-century literature are also found in Curtin *The Image of Africa* (1964), Barker *The African Link* (1978) and Adas *Machines as the Measure of Men* (1989).

6 Although for the sake of convenience I refer to the thousands of miles of coastline as west Africa, Davies (1957) rightly points out that 'Neither politically or economically was West Africa a single unit; the term "African trade" in the later seventeenth century was generic, embracing regions which, because of their divergences, demanded distinct and appropriate commercial techniques' (213). Snelgrave and Atkins both emphasize how very different people were along the trading coasts and that they had extremely varied demands for English goods.

7 Mostly these figures were men, but not always. For the tenuous position of Europeans in Africa, see Davies (1957, chapter 6), Davis (1966, 181–2) and Morgan (1991, 182ff). On the importance of colonies to shaping British opinions about Africans and their difference from Europeans, see Barker (1978, 20–1) and Morgan (1991 213–15); for example, 'The best-developed plantation societies in the New World were the sources of the most virulent comments about blacks' (214). Moreover, even though many of these writers were partici-

pants in the slave trade, 'their attitudes, both to slavery and to Negro culture, were not monolithic' (Barker 1978, 18). Concurring with Barker's characterization of the early eighteenth century, Pratt argues that later readers of travel narratives also 'received nothing like a fixed set of differences that normalized self and Other in fixed ways' (1985, 160). Marshall and Williams (1982) agree with Barker that these earlier accounts do not depict Africans as subhuman, but they do not accept his related contention that 'the Negro image was little different from that of the rest of the uncivilized world' (Barker 1978, 120). Marshall and Williams suggest that in comparison to contemporaneous narratives of other non-Europeans, accounts of Africa possess a 'sharper, more denigratory tone of language' because the writers' preoccupation was the slave trade (229). My own position is that although there was no uniform depiction of Africans, but contradictory stories, there was a dominant view that even detractors such as Atkins and Moore were aware of and worked against.

 8 These frustrations were, of course, offset by the potential for profit: 'In 1700 The Royal African Company expected to sell slaves at four times the value of trade goods paid for them, while private traders expected a return of six to one' (Wolf 1982, 198).

 9 Hallett's introduction to *Records of the African Association* (1964) addresses this issue in a different way than I do. He argues that because trade in gold became less important than slaves, and because the slave trade was so well organized on the coast, 'There was therefore no incentive for European traders to visit the unknown parts of the continent' (7). While this may be so, there was such a consistent mentioning of their inability to know the hinterland that I interpret this common-place as revealing more than practical difficulties.

10 Earlier in the century, Swift used the analogy of European knowledge about Africa to critique contemporary poets. Although he exaggerates, there is a ring of truth to it: 'So Geographers, in *Afric*-maps/With Savage-pictures fill their Gaps, /And o'er uninhabitable Downs/Place Elephants, for want of Towns (Swift 1733, 'On Poetry: A Rhapsody' ll. 176–80).

11 These names were not entirely accurate in terms of the commodities actually available in certain coastal areas; for example, Smith expostulates, 'Why this is call'd the *Gold* Coast I know not? Seeing, that other Countries in *Guinea*, produce as much or more and better Gold, especially about *Gambia*' (1744, 138).

12 Ogilby's maps offer a similar sense of detail and absence. For example, ostriches appear in Nubia and elephants by the subequinoctial circle near Biafara.

13 Bartholomew Stibbs, Edward Drummond and Richard Hull, all employees of the Royal African Company, published *Journal of a Voyage up the Gambia* in 1723. Also see Smith (1744, 24) for references to inland exploration.

14 Some contemporary travellers gave no credence to rumours of cannibalism and felt it necessary to dispute such assumptions. The significant point is that there was no assurance, simply further speculation at this time. Atkins's critique is useful for analysing the ideology of cannibalism: 'My Aim, therefore, was to shew in the best manner I could, that the Accusation every where has probably proceeded from Fear in some, to magnify the Miracle of escaping an inhospitable and strange Country, and from Design in others, to justify Dispossession, and arm Colonies with Union and Courage against the supposed enemies of Mankind. Conquest and Cruelty, by that means go on with pleasure on the People's side, who are persuaded they are only subduing of brutish Nature, and exchanging, for their mutual Good, SPIRITUAL for TEMPORAL INHERITANCES. By particular and private Men, this may have been fixed on a People, to allay some base or villainous Actions of their own, that could not any other way be excused' (Atkins 1735, xxiii). Although some narratives denied cannibalism's existence, geographies commonly included it as part of the African landscape.

15 Trading was supposed to make Britons more civilized too; the very action of contacting and negotiating with foreigners helped make a people civil.

16 Snelgrave's preface and extensive introduction contain no pagination, though the text itself does. All quotations without page references come from the introduction.

Even where Europeans had an established presence, their influence could not obtain the desired information. For example, along the gold coast there were many Dutch and English factories, each with a 'negroe town' attached to it which catered to its needs. Snelgrave comments on the limits of English influence in the areas surrounding their forts: 'The Reader may reasonably suppose, that here we might have a perfect account of the Inland parts; but we can have no such thing'. In fact, Snelgrave laments at each factory that the English are forbidden access to and knowledge about the inland.

17 Mary Louise Pratt and Barbara Stafford are both interested in the scientific and aesthetic forces shaping what travellers 'see' and write, especially in the later eighteenth and nineteenth centuries. In particular, Pratt's *Imperial Eyes* (1992) offers a superb narrative analysis of pre- and post-Linnaean travel literature. She connects the emergence of the scientific explorer with an investigation of the 'anti-conquest', a new paradigm in European travel literature.

18 Atkins's observation is typical of this tendency when he discusses stops from Cape Coast to Whydah where the private traders restock before departing for the West Indies: 'At *Montford, Shallo,* and thereabouts, they make up the Deficiency of Rice and Corn for the voyage, the Country appearing fruitful, and with better Aspect than any of those we have passed to Windward, intermixed with Hills and Vales; at every League almost, a Town; many Corn-fields, Salt pans and other Marks of Industry' (1735, 107).

Very occasionally, as in Francis Moore's *Travels* (1738), the beauty of the land is interrupted by the realities of trade. Moore's journey up the Gambia to assume the position of writer to the Royal African Company at James Fort is pleasurable because the country looked beautiful, woody (i.e. full of natural resources) and cultivated (i.e. a sign of a civilized and industrious people); his pleasure is abated by the fact that this stretch is populated with kings demanding payments from the Europeans who trade there.

19 Popular British accounts often contained significant sections from Bosman's text, a common practice in early travel narratives (Barker 1978, 17). For instance, although Moore challenges his claim about mining, Smith confirms it.

20 Although some writers accuse Africans of an underdeveloped work ethic (see Coetzee 1988, chapter 1), others praise their cultivation skills. Bosman, however, seems determined to find fault with either lack or competency of African labour. He admires the fruitfulness of the Guinea coast, but he condemns the inhabitants for too aggressively planting the land: 'For the Negroes of this Country are so covetous, that no place which is throughout fertile can escape planting' (1705, 339).

21 Pratt's insights into narrative treatment of land and space in the early eighteenth century are drawn from James Turner's work on landscape in the seventeenth century. Early landscape is not a particularized portrait as much as an ideal construction (1992, 45, 48).

The domestic counterpart of the 'improving eye' was also an integral part of a new capitalist ideology. The patriotic endeavours that complemented imperialist expansion abroad included improving the English countryside by building roads, bridges, canals and other infrastructure to permit rapid transportation of people and commerce (Williams 1973, 60–7).

22 Winthrop Jordan's 'First impressions: initial English confrontation with Africans' in *White over Black* (1968) is one of the most helpful studies because it provides

a broad frame of reference and expectations which the English inherited from Classical sources and from other Europeans.

23 Ronald Meek's *Social Science and the Ignoble Savage* (1976) traces the rise, development and uses of four-stages theory in the eighteenth century, especially the way that Native Americans were used as exemplary savages in philosophical discourse.

Although Henry Louis Gates, Jr argues for the seventeenth-century origins of the emergence of arts and letters (literacy) as the sign of civilization that excluded Africans from the human community, the bulk of his evidence is post-1760 in chapters 1 and 2 of *Figures in Black* (1985) and in chapter 4 of *Signifying Monkey* (1988). Thus, while he locates traces earlier, I understand the importance of his research to be identifying a dominant mode of the later period in which the central preoccupation is slavery, not race *per se*.

24 In four-stages theory, shepherd societies were superseded by agricultural ones; trade and commerce crown the hierarchy.

25 Browne's contemporary Matthew Hale (1677) argues that human beauty consists 'principally in these things, figure, symmetry, and colour' (64). The lighter the colour, the more beautiful to Hale: 'in the torrid Climates the common colour is black or swarthy, yet the natural colour of the temperate Climates is more transparent and beautiful' (65). Such an understanding of white complexion as beautiful is certainly commonplace but not universal.

26 Later, both Hogarth and Reynolds draw on notions from travel accounts and natural history about colour, taste, aesthetic judgment; see Hogarth (1753, 188) and Reynolds (1797, 94, 110, 137, 233).

27 Bradley's division begins with white, bearded Europeans and progressively separates populations as they vary from this model; the second sort is white men in America with no beards; copper-coloured Mulattos with straight black hair is the third; Blacks with straight black hair is the fourth; and lastly the Blacks of Guinea with curly black hair, who represent the most differences from Europeans (1721, 169).

Literary historians of the English Renaissance have argued the symbolic currency of blackness in visual spectacles such as Ben Jonson's court masques and in the Lord Mayor pageants in the seventeenth century (Barthelemey 1987, chapter 2). This work has tended to suggest that blackness signified more stably than I am claiming in textual rather than visual productions. See essays by Loomba, Boose and Singh in *Women, 'Race', and Writing in the Early Modern Period* (Hendricks and Parker 1994) for the most recent work on the signification of blackness in the Renaissance spectacle.

28 A typical expression of African difference from Europeans in the 1720s is Bulfinch Lamb's comment about King Trudo Audati of Dahomey: '"and though he seems to be a Man of great natural Parts and Sense as any of his Colour, yet he takes great Delight in trifling Toys and Whims"' (Smith 1744, 182). Some identified colour as the greatest difference quite early in the century: 'The black Colour, and woolly Tegument of these *Guineans*, is what first obtrudes it self on our Observation, and distinguishes them from the rest of Mankind, who no where else, in the warmest Latitudes, are seen thus totally changed' (Atkins 1735, 39).

29 Contemporary narratives explain nakedness in a particular fashion. Most first claim that Africans are naked, then they qualify that statement by stating that only male and female torsos are bare. Finally, they all note when men and women are fully clothed – a sign of their high rank. Thus, the conventions of African society equated rank with the quantity of clothing, though on a rather different scale from the English.

30 See Akinjogbin's *Dahomey and Its Neighbours 1708–1818* (1967) for a useful modern interpretation of the political history.

31 Snelgrave's response to human sacrifice pits his religion as a clear boundary between him and his hosts and at the same time diverts questions about the slave trade, if not justifies it. Snelgrave watches a sacrificial beheading and is asked how he likes the sight. Snelgrave replies: "'Not at all: For our God had expressly forbid us using Mankind in so cruel a manner '" (1734, 45).

32 Peter Earle has also observed that 'it is part of Captain Singleton's mission to demonstrate to Defoe's readers the possibilities and gains to be had from Africa' (Earle 1977, 54).

33 In fact, as Penelope Wilson notes in her introduction to the recent Oxford University Press edition, until the 1780s, *Captain Singleton* was not advertised as a fictional work by Defoe (1720, vii); it thus circulated as a 'true', if sensational, memoir. The title page accentuates the travel book quality and builds interest especially around Africa. The title page bills Africa as the most exciting part of the entire novel: 'the Life, Adventures, and Pyracies, of the Famous Captain Singleton: Containing an Account of his being set on Shore in the Island of *Madagascar*, his Settlement there, with a Description of the Place and Inhabitants: Of his Passage from thence in a Paraguay, to the main Land of *Africa*, with an Account of the customs and Manners of the People: His great Deliverances from the barbarous Natives and wild Beasts: Of his meeting with an *Englishman*, a Citizen of *London*, among the *Indians*, the great Riches he acquired, and his Voyage Home to *England*: As also Captain *Singleton's* Return to Sea, with an Account of his many Adventures and Pyracies with the famous Captain *Avery* and others'. Secord's study (1924) looks at travel narratives Defoe was familiar with and that were in his personal library. Scrimgeour's examination is also valuable; his conclusion is that 'he does not exert himself to use what information was available' (1963–4, 29).

34 At the time of the novel's publication or, indeed, one hundred years later, few Britons could have questioned the accuracy of Defoe's depiction. Many literary critics have remarked on *Captain Singleton*'s geography and its representation. My interest is in the historical and economic significance of that geography to English imperialism and the representation of that interest.

35 See Secord (1924, 129–30) and Knox-Shaw 'Defoe's wilderness: the image of Africa in *Captain Singleton*' (1987) for the state of maps of Africa's interior. Interestingly, several scholars have even remarked upon the prescient geographical knowledge Defoe shows, especially his nineteenth-century critics. In a typical comment of the time, essayist Charles Lamb writes: 'I have traced this route on the map of Africa, in Wyld's great Atlas of 1849; and where I find the map *carte blanche* for more than a thousand miles, Defoe's Captain Singleton, of 1720, has guided me along the shores of the mighty Lakes . . . through dreary deserts, and across primeval mountains. . . . This knowledge of the interior of Africa appears the more amazing since the recent researches of Livingstone, Baker, Grant, and other explorers, have confirmed what our author had so long before stated' (Rogers 1972, 180).

36 Novak contends that most of Defoe's fiction concerns 'parts of the world he believed might be profitably colonized . . . [his fiction is] economic propaganda for the planting of new English colonies and the continued development of those already established in North America and the West Indies' (1962, 140). Although I acknowledge the general spirit of Novak's observation, I question the unproblematic correspondence he draws between Defoe's fiction and non-fiction.

37 See *A General History of Africa*, vol. 5, chapter 28. In this chapter, Kent refers to two attempts at English colonization, one of them by a group of Puritans that was sent to Madagascar in 1645 to establish a colony (1992, 863).

38 Arthur Secord's 1924 study was one of the first to observe this phenomenon: 'Singleton's remarks concerning the people and the country are cautiously general' (128).

39 Atkins, no defender of slave trading, though assured of European superiority, reveals the same contradictory statements (1735, 86, 99, 106). Notions of civility appear in eighteenth-century domestic fiction as 'courteous', 'good natured', or having 'genteel manners' (most often associated with men); in a colonial context, civility takes on the added qualities of a good trader and bravery.

40 Although not as attractive as some other items of trade, coins were imported for ornamentation. In the novel, however, the value-added labour is European not African. 'The Royal African Company also imported coins into the Gambia. . . . Coins were used as materials for making ornaments such as bracelets' (Rodney 1970, 171). Eighteenth-century travellers observed that ornamentation was popular among Africans: some paint their bodies and faces, others scarify them. Atkins notes that on the grain and malaguetta coasts, the men wear bracelets on wrists and ankles, fingers, toes, and necklaces (1735, 61).

41 Iron 'was a rare instance of a European import that could be put to productive use by Africans . . . it went into weapons and agricultural implements' (Rodney 1970, 184). Since the 1680s, England had exported a large amount of iron to enhance trade in Upper Guinea (Ibid., 185).

Laura Brown argues that when gold is at issue, as it is in the quotation cited in the text, Africans receive different discursive treatment: 'On these occasions, they are not treated as trees or beasts, but as human beings basically comparable, if radically inferior to the Europeans' (1993, 163).

42 *A New Collection of Voyages and Travels* (1745–7) reprints many sixteenth-century narratives to Africa. What is remarkable are the lists of African commodities for which the English traded; there is no mention made of slaves, although Towerson's 1555 narrative notes the Africans strongly resent any taking of African men for slaves.

43 See Bosman (1705, 52, 82, 334) and Snelgrave (1734, 2).

44 In contrast, the Portuguese foresaw the possible erosion of their local power with the dissemination of European technology and officially refused to trade in weapons (Rodney 1970, 175). 'Between 1703 and 1719, the [Royal African] Company requisitioned a large variety of firearms for the African trade' (Ibid., 175), though arms did not penetrate inland as easily (Ibid., 176–7); guns and other European weapons were not as significant in the early part of the eighteenth century as they were later as a major British export to Africa.

45 The possibility of Africans being cannibals is dismissed early on in the novel: 'He told me he was afraid I would have little Need of Clothes, and that he had been told that the Inhabitants were *Cannibals* or *Men-eaters* (tho' he had no Reason for that Suggestion)' (13), and religion is mentioned only once (58). In this one instance, Christianity does carry significance when Singleton's isolated meditation on his absolute difference from Africans converges with his first religious sentiment. These two feelings occur at a moment of supreme power over the Africans he has helped enslave. He promises the Africans he will not kill them, he will feed them, and protect them from wild animals; to assure the Africans of his intentions, the prince urges Singleton to clap his hands toward the sun. The response of the Africans to his sign of power and fidelity leads him to reflect:

> I think it was the first time in my Life that ever any religious Thought affected me; but I could not refrain some Reflections, and almost Tears, in considering how happy it was, that I was not born among such Creatures as these, and was not so stupidly ignorant and barbarous: But this soon went off again, and I was not troubled again with any Qualms of that Sort for a long time after.
>
> (Defoe 1720, 61)

Singleton's most profound religious insight occurs at the moment when the

Africans submit to him. He employs religious feeling to distinguish himself from the Africans even as he imitates their religious ceremony.

46 Winthrop Jordan also observes this phenomenon: 'The condition of savagery – the failure to be civilized – set Negroes apart from Englishmen in an ill-defined but crucial fashion' (1968, 24). See Hulme's *Colonial Encounters* (1986) on savagery and its (mis)uses, especially 20ff.

47 Other critics have mentioned that nakedness forms a crucial sign of non-Europeanness: 'Singleton makes much of the superficial differences between himself and the Africans, repeatedly mentioning, for instance, their nakedness' (Blackburn 1978, 125). Sutherland also develops this line of argument (1971, 145). An instructive companion piece to this repeated portrayal of African nakedness that reveals one of the economic underpinnings of such a representation is Defoe's paean to domestic clothing production in *The Complete English Tradesman* (1726) in which he includes a 'picture of the influence of inland trade by enumerating the garments which the poorest countryman wore, indicating the region of England from which each article was bought' (Moore 1958, 318).

48 Re-exports were increasingly important to the British and global economies. Cloth, especially East India cottons, calicoes and prints, 'commanded a ready sale in Africa' (Davies 1957, 170). English cloth played a significant role as well in the later seventeenth century; domestic manufacture completely replaced cloth bought from Amsterdam for sale in Africa at this time (Ibid., 175). English woollens were the most important domestic manufacture shipped to Africa (Ibid., 176–7).

49 Another possibility is that African nakedness (which almost always involves some covering, if scanty by European standards) offends the commercial desires underwriting so much of the narrative. Africans don't consume enough!

50 For example, in *An Essay Upon the Trade to Africa* (1711), Defoe defends the necessity and viability of the Royal African Company, a minority position in the eighteenth century (Davies 1957, 130, 150). Defoe refers repeatedly to the continuation of the slave trade under the Company's direction as benefiting the nation (1711, 15). See *Review* I. 44 (1713) for Defoe's assigning mutual blame among the Company, Government and free traders (Vol. V, 557), but see his more usual position in I. 45 (1713) and Vol. IX, 89–90 for discussions of the company versus the free traders.

Defoe not only wrote about but invested in the Royal African Company and the South Sea Company (Moore 1958, 86). He invested £800 for two shares 'which fell so in price that he got less than £100 for them. By June, 1710, he had no investment in the [Royal African] company' (Ibid., 289). Moore adds that Defoe's views on slavery 'are not basically inconsistent. He pointed out that the slave trade was profitable. . . . But when he spoke of the moral wrong, he was more than a century in advance of his fellow countrymen' (289). This cover story for Defoe is generally rendered in a less gentle way. Earle, for example, suggests a different emphasis: Defoe's views of slavery took second place to his interest in economic gain (70).

51 For reference to Africans as prisoners, see 54, 56, 60, 63–5, 78, 80.

52 Contemporaries noted that prisoners were increased because the Europeans inflamed historical tribal differences within Africa and created new ones to increase the flow of slaves. As Davies notes, an increase in slaves usually meant a related decline in crops, and food shortages resulted since burning was a major technique of African warfare (1957). Also consult Davis's study of slavery for information about procurement of slaves for sale to Europeans: 'As late as 1721 the Royal African Company asked its agents to investigate the modes of enslavement in the interior and to discover whether there was any source besides "that of being taken Prisoners in war time?"' (1966, 183).

53 Atkins relates an extremely revealing practice among the English slavers: assigning titles of 'lord', 'duchess' and even 'prince' to the children of local potentates. 'To smooth the King [of Sesthos] into a good Opinion of our Generosity, we made it up to his Son, *Tom Freeman*; who, to shew his good-nature, came on board uninvited, bringing his Flagelet, and obliging us with some wild Notes. Him we dress'd with an edg'd Hat, a Wig, and a Sword, and gave a patent upon a large Sheet of Parchment, creating him Duke of Sesthos' (1735, 66). Again, clothes make the man.

54 I am not claiming that there were no racialized factors that meant African people rather than Chinese, for instance, were enslaved. Although this cause and effect relation has been widely discussed in fifty years of response to Eric Williams's seminal research on the economic bases of slavery, I believe it is neither as simple as the claim that racism preceded slavery (and hence caused it) or that racism emerged with the threatened abolition of slavery (and hence resulted from it).

REFERENCES

Adas, Michael (1989) *Machines as the Measure of Men: Science, technology, and ideologies of Western dominance*, Ithaca: Cornell University Press.

Akinjogbin, I.A. (1967) *Dahomey and Its Neighbours 1708–1818*, Cambridge: Cambridge University Press.

Appiah, Kwame Anthony (1992) *In My Father's House: Africa in the philosophy of culture*, New York: Oxford University Press.

Atkins, John (1735 [1970]) *A Voyage to Guinea, Brazil, & the West Indies*, London: Frank Cass.

Backsheider, Paula (1989) *Daniel Defoe; His Life*, Baltimore: Johns Hopkins University Press.

Barbot, John (1699) *A Description of the Coasts of North and South Guinea*, Churchill V. 1–668.

Barker, Anthony (1978) *The African Link: British attitudes to the negro in the era of the Atlantic slave trade, 1550–1807*, London: Frank Cass.

Barthelemy, Anthony (1987) *Black Face, Maligned Race: The representation of blacks in English drama from Shakespeare to Southerne*, Baton Rouge: Louisiana State University Press.

Blackburn, Timothy (1978) 'The coherence of Defoe's Captain Singleton', *The Huntington Library Quarterly* XLI.2, pp. 119–36.

Boothby, Richard (1646) *A Brief Discovery or Description of the Most Famous Island of Madagascar, or St. Laurence. In Asia unto the East Indies*, in *A Collection of Voyages and Travels*. comp. Thomas Osborne, London, 1745.

Bosman, William (1705) *A New and Accurate Description of the Coast of Guinea. Divided into the Gold, the Slave, and the Ivory Coasts*, transl. London: J. Knapton *et al*.

Bradley, Richard (1721) *A Philosophical Account of the Works of Nature*, London: W. Mears.

Brantlinger, Patrick (1985) 'Victorians and Africans: the genealogy of the myth of the dark continent', in Henry Louis Gates, Jr (ed.), *'Race,' Writing, and Difference*, Chicago: University of Chicago Press, pp. 184–222.

Brown, Laura (1993) *Ends of Empire: Women and ideology in early eighteenth-century English literature*, Ithaca: Cornell University Press.

Browne, Sir Thomas (1646 [1981]) 'Of the Blacknesse of Negroes', in *Pseudodoxia Epidemica; or, Enquiries into very many Received Tenents and Commonly Presumed Truths*, ed. Robin Robbins, Oxford: The Clarendon Press.

Burg, B.R. (1983) *Sodomy and the Perception of Evil: English sea rovers in the seventeenth-century Caribbean*, New York: New York University Press.

Churchill, Awnsham and John (1732) *A Collection of Voyages and Travels*, vol. V, London.

Coetzee, J.M. (1988) *White Writing: On the culture of letters in South Africa*, New Haven: Yale University Press.

The Compleat Geographer (1709) 3rd edition, Awnsham and John Churchill (eds).

Curtin, Philip (1964) *The Image of Africa: British ideas and action, 1780–1850*, Madison: University of Wisconsin Press.

Davies, K.G. (1957) *The Royal African Company*, London: Longmans, Green.

Davis, David Brion (1966) *The Problem of Slavery in Western Culture*, New York and London: Oxford University Press.

Defoe, Daniel (1709 [1938]) *Review* V. 140, Vol. VI, Facsimile Rpt, New York: Columbia University Press.

—— (1711) *An Essay Upon the Trade to Africa, in Order To Set the Merits of that Cause in a True Light and Bring the Disputes between the African Company and the Separate Traders into a Narrower Compass*, London.

—— (1713 [1938]) *Review* I. 44, Vol. IX, Facsimile Rpt, New York: Columbia University Press.

—— (1713 [1938]) *Review* I. 45, Vol. IX, Facsimile Rpt, New York: Columbia University Press.

—— (1720 [1990]) *Captain Singleton*, ed. Shiv Kumar and introduced by Penelope Wilson, Oxford: Oxford University Press.

—— (1726) *The Complete English Tradesman*, London: Charles Rivington.

Duncan, James (1993) 'Sites of representation: place, time and the discourse of the Other', in James Duncan and David Ley (eds), *Place/Culture/Representation*, London: Routledge.

Earle, Peter (1977) *The World of Defoe*, New York: Atheneum.

Gates, Henry Louis, Jr (1985) *Figures in Black: Words, signs, and the 'racial' self*, New York: Oxford University Press.

—— (1988) *The Signifying Monkey: A theory of Afro-American literary criticism*, New York: Oxford University Press.

Hale, Matthew (1677) *The Primitive Origination of Mankind, Considered and Examined According to the Light of Nature*, London: William Godbid.

Hallett, Robin (ed.) (1964) *Records of the African Association 1788–1831*, London: Thomas Nelson.

—— (1965) *The Penetration of Africa: European exploration in North and West Africa to 1815*, New York: Frederick A. Praeger.

Hendricks, Margo and Parker, Patricia (eds) (1994) *Women, 'Race', and Writing in the Early Modern Period*, New York: Routledge.

Hogarth, William (1753 [1908]) *The Analysis of Beauty*, Chicago: Reily and Lee.

Houstoun, James (1725) *Some New and Accurate Observations Geographical, Natural and Historical. Containing a True and Impartial Account of the Situation, Product, and Natural History of the Coast of Guinea, so far as Relates to the Improvement of that Trade, for the Advantage of Great Britain in General, and the Royal African Company in Particular*, London: J. Peele.

Hulme, Peter (1986) *Colonial Encounters: Europe and the native Caribbean, 1492–1797*, New York: Routledge.

Jablow, Alta (1963) The development of the image of Africa in British popular literature, 1530–1910, Ph.D. dissertation, Columbia University, New York.

Jordan, Winthrop (1968 [1977]) *White over Black: American attitudes toward the negro, 1550–1812*, New York: W.W. Norton.

Kent, R.K. (1992) 'Madagascar and the islands of the Indian Ocean', in *General History of Africa: Africa from the sixteenth to the eighteenth century*, Vol. V., ed. B.A. Ogot, Berkeley: University of California Press.

Knox-Shaw, Peter (1987) *The Explorer in English Fiction*, London: Macmillan.

Marshall, P.J. and Williams, Glyndwr (1982) *The Great Map of Mankind: Perceptions of new worlds in the age of Enlightenment*, Cambridge, MA: Harvard University Press.

Meek, Ronald (1976) *Social Science and the Ignoble Savage*, Cambridge: Cambridge University Press.

Moore, Francis (1738) *Travels into the Inland Parts of Africa*, London: Edward Cave.

Moore, John (1958) *Daniel Defoe: Citizen of the modern world*, Chicago: University of Chicago Press.

Morden, Robert (1693) *Geography Rectified; or, a Description of the World*, London: Robert Morden and Thomas Cockerill.

Morgan, Philip (1991) 'British encounters with Africans and African-Americans, circa 1600–1780', in *Strangers within the Realm: Cultural margins of the first British Empire*, Chapel Hill: University of North Carolina Press.

Mudimbe, V.Y. (1988) *The Invention of Africa: Gnosis, philosophy, and the order of knowledge*, Bloomington: Indiana University Press.

A New Collection of Voyages and Travels (1745–7) collected by Green, 2 vols, London: Thomas Astley.

Novak, Maximillian (1962) *Economics and the Fiction of Daniel Defoe*, New York: Russell and Russell.

Nussbaum, Felicity (1994) 'The other woman: polygamy, *Pamela*, and the prerogative of empire', in Margo Hendricks and Patricia Parker (eds), *Women, 'Race', and Writing in the Early Modern Period*, New York: Routledge.

Ogilby, John (1670) *Africa, Being an Accurate Description . . . .*, London.

The Philosophical Transactions of the Royal Society of London (1667) Vol. I.

Popkin, Richard (1973) 'The philosophical basis of eighteenth-century racism', in *Studies in Eighteenth-Century Culture*, Vol. 3, *Racism in the Eighteenth Century*, ed. Harold E. Paglioaro, Cleveland: The Press of Case Western Reserve University, pp. 245–62.

Pratt, Mary Louise (1985) 'Scratches on the face of the country; or, what Mr. Barrow saw in the land of the Bushmen', in Henry Louis Gates, Jr (ed.), *'Race,' Writing, and Difference*, Chicago: University of Chicago Press, pp. 138–62.

—— (1992) *Imperial Eyes: Travel writing and transculturation*, New York: Routledge.

Proceedings of the Association for Promoting the Discovery of the Interior Parts of Africa, (1790) London: C. Macrae.

Rediker, Marcus (1987) *Between the Devil and the Deep Blue Sea: Merchant seamen, pirates, and the Anglo-American maritime world, 1700–1750*, Cambridge: Cambridge University Press.

Reynolds, Joshua (1797 [1959]) *Discourses on Art*, ed. Robert Wark, San Marino, CA: Huntington Library.

Richetti, John (1975) *Defoe's Narratives: Situations and Structures*, Oxford: The Clarendon Press.

Rodney, Walter (1970) *A History of the Upper Guinea Coast, 1545–1800*, Oxford: The Clarendon Press.

Rogers, Pat (1972) *Defoe: The critical heritage*, London: Routledge and Kegan Paul.

Schiebinger, Londa (1990) 'The anatomy of difference: race and sex in eighteenth-century science', *Eighteenth-Century Studies* 23.4, pp. 387–405.

Scrimgeour, Gary (1963–4) 'The problem of realism in Defoe's *Captain Singleton*', *The Huntington Library Quarterly* XXVII, pp. 21–38.

Secord, Arthur (1924 [1963]) *Studies in the Narrative Method of Defoe*, New York: Russell and Russell.

Smith, William (1744 [1967]) *A New Voyage to Guinea*, London: Frank Cass.

Snelgrave, William (1734 [1971]) *A New Account of Some Parts of Guinea and the Slave-Trade*, London: Frank Cass.

Stafford, Barbara (1984) *Voyage into Substance: Art, science, nature and the illustrated travel account, 1760–1840*, Boston: MIT Press.

Stibbs, Bartholemew *et al.* (1723) *Journal of a Voyage up the Gambia*. Published in the same volume with Francis Moore (1738) *Travels into the Inland Parts of Africa*, London: Edward Cave.

Sutherland, James (1971) *Daniel Defoe: A critical study*, Cambridge, MA: Harvard University Press.

Swift, Jonathan (1733 [1969]) 'On Poetry: A Rhapsody', *Eighteenth-Century English Literature*, ed. Geoffrey Tillotson *et al.*, New York: Harcourt, Brace and World.

Watson, Francis (1952 [1969]) *Daniel Defoe*, New York: Kennikat Press.

Williams, Raymond (1973) *The Country and the City*, New York: Oxford University Press.

Wolf, Eric (1982) *Europe and the People without History*, Berkeley: University of California Press.

3 Enlightenment Travels

The making of epiphany in Tibet

Laurie Hovell McMillin

On the day before I flew from Chengdu to Lhasa in 1985, I made the following account in my journal. 'I take with me: 1 down vest; 1 *What the Buddha Taught*; prayer beads given by a rinpoche; 1 Walkman; 1 jar of Nescafé; warm socks; my past karma; a Tibetan vocabulary for hello, turquoise, silver, maroon, yellow, thank you, that's good, that's bad; impressions from books I've read and half-read; a ticket out.

As a longtime reader of accounts of travel to Tibet, I was (painfully) aware of the baggage I carried as a Western traveller there. From James Hilton's *Lost Horizon* to Peter Matthiessen's *The Snow Leopard*, something significant was supposed to happen to Western travellers in Tibetan lands. In the following analysis, I trace out the making of the myth that travel to Tibet might bring on a kind of spiritual transformation. Most broadly, this myth-making is part of a larger British project to make sense of selves and others during the colonial era; more specifically, it is intimately connected to how British scholars and travellers imagine religion for themselves and how they perceive the religion of the Tibetans.[1]

The text in which this myth of transformation in Tibet is fully fleshed out is a 1910 travel account by Francis Younghusband called *India and Tibet*. Younghusband's text draws on various texts that have gone before it, to produce finally a sort of paradigm of transformation with which subsequent accounts of travel to Tibet in English must grapple. And what is striking about Younghusband is that he is not a proto-typical guru seeker; neither is he in Tibet to explore, to write about Tibetan religion, nor even to measure or scale the mountains as many Europeans before him did and continue to do. Younghusband travelled to Tibet with 8,000 Indian Army troops: he was in Tibet to negotiate a treaty with the Thirteenth Dalai Lama and to establish British military and political might in the region at a time when Russia also had designs in Central Asia and Chinese rule in Tibet was on the wane. It is in Younghusband's text – a 450-page document that recounts the tedious negotiations, polemicizes against the sinister qualities of much of Tibetan religion and clergy, and seeks a justification for British imperialism in Asia – that the first instance of what I call a Tibetan epiphany appears: a moment of spiritual and personal realization and revelation which gives Younghusband a new sense of the world and his place in it. Where does this 'revelation' come from? How does Younghusband get there?

Many recent analyses of travel writing have focused on the textual nature of the accounts: the discourses that shape the narrative, the conventions by which it is shaped, the ways in which travel writing is always a re-presentation of an experience that is something other than the account itself. Accounts of Tibetan travel can also be examined fruitfully in this way. The making of Tibetan epiphany is certainly an intertextual phenomenon: travel accounts influence later travel accounts; other genres of writing connect to accounts of Tibetan travel; and certain tropes and conventions slip the borders from biography and poetry to novel such that the epiphany finally winds up in a travel account. The making of Tibetan epiphany is also a textual phenomenon in that it depends upon the struggles individual writers and travellers undergo in accounting for their journeys. But it is also essential that these writers apprehend the place they write about as the same place. They 'write' Tibet and in so doing mobilize existing knowledge on that place. Thus, an analysis that looked only for the operation of colonial discourse or imperialist ideology in these texts would likely miss something about their specificity. For central to almost any discussion of Tibet – from the eighteenth century till today – is religion.

While the estimations of Tibetan religion shift over the years, describing and judging the faith of Tibetans – which Europeans did not confidently recognize as 'Buddhism' until the nineteenth century – seems a necessity. As European knowledge on Buddhism and its Tibetan forms increases, the evaluations of individual travellers move to accommodate these changes, as we will see. And especially striking about the making of the myth of Tibet as a site of non-denominational spirituality is that the key texts in this process are those with an expressly secular orientation towards religion: British writers of these texts keep their distance from Tibetan religion, which for them is a thing to explain, not to live out. Indeed, these writers tend to be sceptical about organized religion in general: they are inclined instead to be adherents of a kind of secular humanist faith that has its roots in European Enlightenment thinking. While we can see the development of the myth of transformation in Tibet as part of a larger de-sacralization of the world that included a refurbishing of conventionally religious concepts for secular life, the continuity of Tibet as a trope and Tibet as a contested political arena must be addressed. Without that concern for the geographical specificity of Tibet as a place written, invaded, surveyed and imagined by Western travellers, the dynamics of the development of the myth of epiphany may well be overlooked. In this study, I explore how travel texts on Tibet are enabled by an 'intertext' on Tibet, paying close attention to the ways in which conceptions of Tibetan religion and the religious positions of the writer contribute to the making of Tibetan epiphany.

BOGLE AND TURNER IN TIBET

Until quite recently, Tibet has historically been either difficult for Westerners to reach or officially off-limits; in the case of Britain, no official delegation representing British interests was allowed to enter Tibet from 1784 to 1904.[2] Tibet's relative inaccessibility had two results: first, those Europeans who knew anything about Tibet seemed compelled to write about it; and secondly, those who were interested in Tibet became especially dependent on written accounts. When George Bogle was sent by Warren Hastings to Xigatse by way of Bhutan in 1774, the first Briton to travel to Tibet was ordered ' "[t]o keep a diary, inserting whatever passes before your observation" ' (Markham 1879, 54) In this journal, Bogle remarks on the paucity of textual sources for his journey and notes on his entry into Bhutan that he was 'in the dark' about the place, as 'the only information I could collect' was '[t]he imperfect account of some religious mendicants' (Ibid., 15).[3] Referring to the accounts of Jesuit and Capuchin missionaries who had travelled in the Himalayas earlier in the century, the representative of the English East India Company is suspicious of these clerics who saw it as their duty to convert the pagans. By way of contrast, Bogle's orientation is largely humanist and relativistic, his purposes political and mercantile. Sent to Tibet to establish commercial and diplomatic relations with the Third Panchen Lama, Bogle seems largely unconcerned with Tibetan religion and notes that '[t]he religion of the Lamas is somehow connected with that of the Hindus, though I will not pretend to say how. . . . The humane maxims of the Hindu faith are taught in Tibet. To deprive any living creature of life is regarded as a crime, and one of the vows taken by the clergy is to that effect' (Ibid., 72). Unlike his Victorian successors, Bogle lacks the conceptual and classificatory apparatus even to name Tibetan religion as a species of Buddhism. For him, then, Buddhism is not Tibet's defining characteristic.

Bogle nonetheless seems compelled to comment on Tibetan religion and in the following passage he offers a tongue-in-cheek explanation of the nature of the differences between the Yellow Caps (Gelukpa) and the Red Caps (Sakyapa and Kagyupa): '[A]s I adhere to the tenets of [the Yellow Caps], and have acquired my knowledge of religion from its votaries, I will not say here much upon the subject. . . . I may be allowed, however, just to mention two things, which must convince every unprejudiced person of the wicked lives and false doctrines of the Red Caps. In the first place, many of the clergy marry; and in the next, they persist, in opposition to religion and common sense, in wearing Red Caps' (Ibid., 180). Bogle waxes ironic on the finer points of sectarian differences, but it is worth noting that he does so by playfully counting himself as one of their number. He can take up this position both because Tibet has not yet been thoroughly inscribed by the West, and because Bogle approaches those he meets with a kind of relativistic humanism, one which looks for the common ground between human beings. This relativistic approach is extended especially to the Panchen Lama, whom he refers to as the Teshu Lama.[4]

When Bogle met Lobsang Palden Yeshe, the Third Panchen Lama, the latter wielded considerable political power in western Tibet and retained some independence from the Chinese-influenced government in Lhasa. Bogle notes that although the Teshu Lama is 'venerated as God's Vice-gerent [*sic*] through all the Eastern countries, endowed with a portion of omniscience, and with many other divine attributes, he throws aside, in conversation, all the awful part of his character, accommodates himself to the weakness of mortals and endeavours to make himself loved rather than feared, and behaves with the greatest affability to everybody, particularly to strangers' (Ibid., 83–4). As one of the strangers on whom the Teshu Lama's affection was bestowed, Bogle dwells less on the supposed divinity of the figure than on his excellent human qualities; Bogle remarks that the Teshu Lama's 'disposition is open, candid, and generous. He is extremely merry and entertaining in conversation, and tells a pleasant story with a great deal of humour and action' (Ibid., 84). Evidently impressed by the man, Bogle seems suspicious of his own opinion and seeks out alternative and critical perspectives. As he writes in his report to Hastings, 'I endeavoured to find out, in his character, those defects which are unseparable from humanity, but he is so universally beloved that I had no success, and not a man could find it in his heart to speak ill of him' (Ibid.).

While Bogle cannot revere the Teshu Lama as a *bodhisattva* – as an Enlightened being in the Tibetan Buddhist sense – he can respect him as an 'enlightened' human in the European sense, as a rational man, a fully developed person. But as a man who thinks relatively about customs, who is willing to imagine other people's views of the world, Bogle at times finds himself in what might seem surprising relations with the lama. In the following passage Bogle describes how the Tibetan leader passes a line of monks and lay followers in a procession.

> As the Lama passed they bent half forwards, and followed him with their eyes. But there was a look of veneration mixed with joy in their countenances which pleased me beyond anything, and was a surer testimony of satisfaction than all the guns in the Tower, and all the odes of Whitehead could have given. One catches affection by sympathy, and I could not help, in some measure, feeling the same emotions with the Lama's votaries.
>
> (Markham 1879, 95)

Carefully explaining his sentiments, Bogle first describes Tibetans in the act of venerating their spiritual and political leader. This act of veneration, which involves gazing, bowing, attending to the figure, is then compared to an English custom and is finally judged more satisfactory. Bogle's relativism allows him to take up the place of the Tibetans here; he sees with what he imagines are their eyes and finds himself sympathetic, finds himself in a seemingly unexpected connection to the lama and to those around him.

Such a relation is possible because religion – whether that of Tibet or that of Europe – is subordinate to Bogle's greater concerns of humanism and mercantilism. In humanism lies the possibility of relation and connection; in mercantilism is the possibility of exchange. Thus when the lama asks Bogle to reconcile the notion of one God with the idea of the Trinity, Bogle feels himself 'unequal' to the task, and assents when the lama surmises that all people worship one god under different names (Ibid., 143). Bogle goes on to argue that he was not sent as a 'missionary . . . armed with beads and crucifixes . . . to convert unbelieving nations' (Ibid.). Both his concern for his diplomatic duty – the secular interests of the East India Company – and his relativistic humanism – a secular ideology – allow him to give questions of religion a subordinate status.

However, if these secular discourses give Bogle a way to think about organized religion, they also finally bring him into conflict. The uneasiness of the alliance between mercantilism and humanism is brought to light at the end of his journey when Bogle expresses his hopes for Tibet's future. In his report to Hastings, Bogle argues that Tibet is ready for mercantile expansion. But in writing to his sister, Bogle offers these sentiments to the country: 'May ye long enjoy that happiness which is denied to more polished nations, and while they are engaged in the endless pursuit of avarice and ambition, defended by your barren mountains, may ye continue to live in peace and contentment, and know no wants but those of nature' (Ibid., 177). By journey's end, Bogle finds himself divided between his duties and his sympathies. He is left with two irreconcilable perspectives. But even if Bogle comes to 'realize' the incommensurability of his duty and his sentiment, this is not yet Tibetan epiphany. Epiphany would involve an encompassing view, a reconciliation of self and world; such transformation will only be possible in later years.

Interests in relativism and mercantilism similarly inform Samuel Turner's account of 1800. Here also a Tibetan lama is treated with respect; here too the relationship finally serves a commercial function. Turner was sent to Xigatse in 1783 to meet the newly reincarnated Fourth Teshu Lama (the previous one had died in 1780 in Beijing). While not as welcome at Tashilhunpo as his predecessor, Turner echoes Bogle's affection for the previous lama and depicts the eighteen-month-old child lama as possessed of extraordinary powers of perception, prediction and intelligence. After Turner offers a speech praising the lama and expressing thanks for his 're-appearance' on earth, he notes that '[t]he little creature turned, looked stedfastly [*sic*] towards me, with the appearance of much attention while I spoke, and nodded with but slow movements of the head, as though he understood and approved every word, but could not utter a reply' (Turner 1800, 334–5). Such intelligence and perception, Turner hopes, will make the infant lama recall his earlier affection for the East India Company. In his report to Hastings, Turner argues his case: 'If the Teshoo Lama shall be made to resume the plans projected by him in his presumed existence, . . . the same consistency of conduct will certainly prompt him to look back to the negociations [*sic*] of 1775, to the free intercourse of trade between Tibet and Bengal, which then coincided with his desires' (Ibid., 366). In a rather functionalist use of relativism towards Tibetan religion, Turner assents to the possibility of reincarnation in order to assert the prospect of trade: 'I ground my hope on presumptions built upon the tenets of their faith, which is the basis on which their government is itself constructed' (Ibid., 378). Such polite participation in local customs works on behalf of the Company; in effect, Turner uses Tibetan religion to further the secular goals of the commercial and political enterprise.

Turner had read Bogle's account in manuscript, and at several points draws on Bogle's precedent to explain his attraction to the young lama (Ibid., 338).[5]

Bogle's text had suggested the strength of a human connection between Tibetan and Briton; this is echoed in Turner and will be reshaped and re-imagined in later texts. Translated to other historical, cultural and ideological moments, the figure of the Tibetan lama will have other work to do.[6]

BUDDHISM DEFINED

Neither Bogle nor Turner could count Tibetan religion as an instance of Buddhism, but the work that would eventually describe, name and classify the thing that most Tibetans practised was already in its formative stages. Under the governorship of Warren Hastings, scholars such as William Jones, Charles Wilkins and H.T. Colebrooke began formal study of Indian religion and language in the late eighteenth century. As part of an Enlightenment project of classifying 'other peoples' religion',[7] the Orientalists located the truth of Indian religion in written documents rather than in current practices; for them, texts provided evidence of a bygone Hindu Golden Age.

The early Orientalists focused on what later came to be known as Brahmanism and Hinduism, but even then Buddhism was an area of interest. Positing similarities between the living religions of Siam, Burma, Ceylon, Tibet and China, European scholars in the early nineteenth century began to classify various practices, beliefs and texts as part of the religion of Buddha or Buddhism (Almond 1988, 13). By the 1830s, Europeans understood Buddhism as defining the religious beliefs and practices of most of Asia. As Philip Almond notes in *The British Discovery of Buddhism*, from the late eighteenth century to the 1830s, Buddhism becomes 'an object' for Western scholars: 'it takes a form as an entity that exists over and against the various cultures which can now be perceived as instancing it, manifesting it, in an enormous variety of ways (Ibid., 12). For the Orientalists, texts were the basis of this entity; texts provided a standard against which diverse practices and customs could be measured. With this value for scripture as the source of authentic, stable and pure Buddhism, British scholars turned to the case of Tibetan religion. Two scholar-officials are of special note: Brian Hodgson and L. Austine Waddell.

The comprehensive nature of Hodgson's work is evident from the title of his collected studies: *Essays on the Languages, Literatures, and Religion of Nepal and Tibet Together with Further Papers on the Geography, Ethnology, and Commerce of those Countries*. Trained at the College of Fort William in Calcutta, Hodgson conducted his researches on Tibet from Nepal, where he acted as Assistant Resident, Postmaster and Resident of Kathmandu between 1820 and 1842. Forbidden to travel to Tibet, Hodgson turned practical difficulty into methodological directive and argued that texts were the best source for the study of Buddhism.[8]

For Hodgson, Sanskrit texts represent the pinnacle of Buddhist thought. He writes: 'Buddhism arose in an age and a country celebrated for literature; and the consequence was, that its disciplines were fixed by means of one of the most perfect languages in the world (Sanskrit), during, or immediately after, the age of its founder' (Hodgson 1972, 99).[9] In a kind of scholarly fundamentalism, Hodgson suggests that later translations and commentaries only distort an original and self-evident perfection. So confident is Hodgson in the superiority of Sanskrit texts that he declares his desire to 'separate Buddhism *as it is* . . . and Buddhism as it ought to be' (Ibid., 41; emphasis in original).

Echoing William Jones's work, Hodgson finds Sanskrit comparable to Greek in that both are 'capable of giving a soul to the objects of sense, and a body to the abstractions of metaphysics' (Ibid., 66). Through Sanskrit, he argues, scholars may discover 'very many things inscrutably hidden from those who were reduced to consult barbarian translations from the most refined and copious of languages upon the most subtle and interminable of topics' (Ibid., 110). Further discrediting the research of Alexander Csoma de Körös, a Czech scholar who dealt with Tibetan language and texts, Hodgson argues that Csoma de Körös's 'attainments in Tibetan lore have been comparatively useless' because he has relied on 'Tibetan translations of my Sanskrit originals' (Ibid., 65, 66).[10] Dubious of the authenticity of vernacular translations, Hodgson notes: 'whoever will duly reflect upon the dark and profound abstractions, and the infinitesimally-multiplied and microscopically distinguished personifications of Buddhism, may well doubt whether the language of *Tibet* does or can adequately sustain the weight that has been laid on it' (Ibid., 66; emphasis in original). For Hodgson, Tibetan texts are 'barbarian translations' (Ibid., 110); indeed, it is remarkable to him 'that literature of any kind should be . . . so widely diffused as to reach persons covered with filth' (Ibid., 9).

Despite Hodgson's conviction that 'documentary is superior to verbal evidence' and his suspicion of those scholars who 'prate about mere local rites and opinions' (Ibid., 100), he maintains that local informants can be of some use. '[W]hatever may be the general intellectual inferiority of the orientals of our day, let us not suppose that the living followers of Buddha cannot be profitably interrogated touching the creed they live and die in' (Ibid.). But this profitability too resides in their association with the texts of Buddhism; living Buddhists guard against the complete degeneration of their religion 'by the possession and use of original scriptures, or of faithful translations of them, which were made in the best age of the church' (Ibid.).

Hodgson participates in a wider Victorian tendency to create an ideal textual Buddhism with which the practices of living Buddhists might be judged. As Almond argues, '[T]hose who saw Buddhism in the East in the second half of the [nineteenth] century could not but measure it against what it was textually said to be, could not but find it wanting and express this in the language of decay, degeneration, and decadence' (Almond 1988, 37). In effect, Hodgson and others like him sacralize the texts that seem to comprise Buddhism, canonizing them: texts become the source of the sacred truth about Buddhism, a truth to which Europe has privileged and seemingly unprejudiced access. Edward Said uses the term 'textual attitude' to describe the Orientalist preference for 'the schematic authority of a text to the disorientations of direct encounters with the human'

(Said 1979, 93). By the middle of the nineteenth century, Western scholarship had thoroughly integrated the textual attitude such that, in Almond's words, 'the Buddhism that existed "out there" was beginning to be judged by a West that alone knew what Buddhism was, is, and ought to be. The essence of Buddhism came to be seen as expressed not "out there" in the Orient, but in the West through the West's control of Buddhism's own textual past' (Almond 1988, 13).[11]

The value for textualized Buddhism is also part of L. Austine Waddell's influential study of 1895. The title of the second edition of this work is telling: *The Buddhism of Tibet or Lamaism* strives to differentiate authentic Buddhism from what Tibetans actually do, which in Waddell's eyes is to (falsely) institutionalize reverence for lamas. Since Tibet was officially off-limits to European travellers, Waddell, like Hodgson, had to conduct his researches in Tibetan religion from border regions, in his case Darjeeling and Sikkim. And while Waddell was more expressly concerned with what Tibetan Buddhists actually practised – going so far as to purchase a monastery where he could undertake his investigations[12] – his work is also influenced by the notion that real Buddhism lies in texts.

On the first page of his study, Waddell makes the following assertion: 'To understand the origin of Lamaism and its place in the Buddhist system, we must recall the leading features of primitive Buddhism, and glance at its growth, to see the points at which the strange creeds and cults crept in, and the gradual crystallization of these into a religion differing widely from the parent system, and opposed in so many ways to the teaching of Buddha' (Waddell 1895, 5). For Waddell, the history of Buddhism is the story of a downward spiral from the Buddha's original lofty teachings.[13] By the time Buddhism reaches Tibet it is an 'impure form . . . covered with foreign accretions and saturated with so much demonology' (Ibid., 29). After this, 'the distorted form of Buddhism introduced into Tibet . . . became still more debased' (Ibid., 15). What present-day scholars might interpret as the inevitable adaptation of Buddhism to existing cultural, political and historical contexts, Waddell is inclined to read as an ugly syncretism: 'Primitive Lamaism may . . . be defined as a priestly mixture of Sivaite [*sic*] mysticism, magic, and Indo-Tibetan demonology, overlaid by a thin varnish of Mahayana Buddhism. And to the present day, Lamaism still retains this character' (Ibid., 30).

In its texts, Tibetan Buddhism still 'preserves much of the loftier philosophy and ethics of the system taught by Buddha himself' (Ibid.), but these admirable aspects are compromised by the practices of the lamas, who jealously keep this knowledge hidden from lay men and women. Waddell contends that the laity has 'fallen under the double ban of menacing demons and despotic priests. So it will be a happy day, indeed, for Tibet when its sturdy over-credulous people are freed from the intolerable tyranny of the Lamas, and delivered from the devils whose ferocity and exacting worship weigh like a nightmare upon all' (Ibid., 573). In this Protestant criticism of the power of priests, Waddell reasserts Europe's authority to preserve and define 'true' Buddhism as a moral if atheistic creed, one with a systematic and ethical if largely pessimistic philosophy.[14] If one only knows how to read them, the texts contain the truth about Buddhism: the world simply fails to measure up.[15]

A PIVOTAL MOMENT

Following on the heels of Waddell's important study is a book which both depends upon written accounts of Tibetan religion and contributes significantly to the making of the myth of travel as transformation in Tibet. Neither a study of Tibetan religion nor another Tibetan travel account, Rudyard Kipling's novel *Kim*, published in 1901, tells the story of an orphan of Irish descent on the loose in India. Although *Kim* is essentially a coming-of-age story situated in colonial India, the novel also depends upon the 'textual attitude' towards Buddhism; it inherits and refashions the figure of the lama; and it plays out an Enlightenment value for a secular, human-centred world. With demonized Buddhism providing intermittent colour to the story and a gentle lama as sidekick, the hero Kim comes ultimately to value the things of this world and the people in it.

To consider first the import of the textual attitude towards Buddhism, the novel describes the religion of some hill people as 'an almost obliterated Buddhism, overlaid with nature-worship fantastic as their own landscape, elaborate as the terracing of their tiny fields' (Kipling 1901, 281). Figuring Tibetan Buddhism in Waddell's demonizing terms, Kipling refers to Tibetan ritual dances (Tibetan: *cham*) as 'devil-dances' and notes the 'horned masks, scowling masks, and masks of idiotic terror' used for such rituals (Ibid., 197). Similarly, *thangkas* representing various fierce deities are 'fiend-embroidered draperies' (Ibid.). To Britons schooled in images of Christ and Mary designed to elicit devotion and reverence, such figures could only signify *worship* of evil forces. All in all, such seemingly demonic representations offered evidence of the ways in which various excrescences had obscured the dignified Buddhism gathered and translated in European texts.

As if acknowledging his sources, Kipling displays a collection of European texts on Buddhism in the opening pages of the novel. In the Lahore museum, a structure the lama takes for a temple of sorts, the curator shows the Tibetan monk 'a mound of books', the physical manifestations of 'the labours of European scholars, who by the help of these and a hundred other documents have identified the Holy Places of Buddhism' (Ibid., 56). Kipling's lama is sufficiently reverent before the eminent pile, and his comments on Tibetan religion further ally him with the European criticism of Tibetan superstition: 'For five – seven – eighteen – forty years it was in my mind that the Old Law was not well followed, being overlaid . . . with devildom, charms, and idolatry' (Ibid.). The lama's estimation of living Buddhism, which corresponds with Europe's view of it, elevates him, as do his disdain for elaborate ritual, his humility and his determination to seek out the truth for himself. But wait – where does this lama come from?

Kipling's lama is called 'Teshoo Lama' (Ibid., 152). In effect, Kipling refashions the figure described by Bogle and Turner – the second highest incarnation in Tibetan Buddhism and a prominent political power – as a more generalized Tibetan lama. And in the relationship between the boy Kim and the lama are echoes of the affection described by Bogle for 'his' Teshu Lama.

Curiously, in all the critical work on *Kim* – and there is plenty these days – I have encountered no mention of the importance of accounts of the historical Teshu/Teshoo Lama (or Lamas) for Kipling's lama. Like Bogle's Teshu Lama, Kipling's lama is dignified, compassionate, kind and affectionate towards his Anglo-friend and towards other living creatures. Kipling's lama calls a snake 'brother' (Ibid., 92) and abhors violence in general. Bogle's lama and Kim's lama both occupy important positions in the Tibetan hierarchy but somehow manage to preserve their modesty. And Kipling's lama, like the Teshu/Teshoo Lamas met by Bogle and Turner, has a certain usefulness in colonial machinations. But most importantly, both lamas are dignified by their humane attributes rather than their supposed divine ones. Thus, the relationships between Bogle and his lama on the one hand and Kipling's lama and Kim on the other are essentially friendships and not master–student relationships, despite Kim's designation as *chela* (disciple). The depth of these friendships is best exemplified in scenes of departure, and it is significant that such a scene will mark the turning point in Younghusband's text.

When Kim leaves the lama to go to St Xavier's school, his sad farewell echoes Bogle's sorrowful departure from Tashilhunpo. Kim cries, 'I have no friend save thee, Holy One. . . . [H]ow shall I ever forget thee?' (Ibid., 170). So attached is Kim to the lama that he almost forgets who he is: ' "But whither shall I send my letters?" wailed Kim, clutching at his robe, all forgetful that he was a Sahib' (Ibid., 171). After reciting a proverb on the illusory nature of desire, the lama replies, ' "Go to the Gates of Learning. Let me see thee go . . . Dost thou love me? Then go, or my heart cracks" ' (Ibid.).

On his departure from the Teshu/Panchen Lama, Bogle too is moved. Writing to his sister, Bogle notes that '[a]s the time of my departure drew near, I found that I should not be able to bid adieu to the Lama without a heavy heart. The kind and hospitable reception he had given me, and the amiable disposition which he possesses, I must confess had attached me to him, and I shall feel a hearty regret at parting' (Markham 1879, 177). In his report to Hastings he writes: 'I never could reconcile myself to taking last leave of anybody, and what from the Lama's pleasant and amiable character, what from the many civilities he had shown me, I could not help being particularly affected. He observed it, and in order to cheer me mentioned his hopes of seeing me again. He threw a handkerchief about my neck, put his hand on my head, and I retired' (Ibid., 171). The affection between Anglo and lama becomes heightened on departure, but in both cases the connection is a deep friendship, a cross-cultural, interracial, homo-social bond, if you will, but not a guru–disciple relationship. From the white character's perspective, the friendship, though approaching the sublime, is a human relationship based on human emotion. Indeed, from the novel *Kim*'s point of view, the lama is deceived in thinking Kim is his *chela*, his disciple, for Kim's devotion to the lama is of a secular type. Without Kim as protector, without the paternal arm of imperialism, the lama, with his pacifist and innocent ways, could hardly survive. As an Indian character in the novel sums up the situation, '[I]f evil men were not now and then slain it would not be a good world for weaponless dreamers' (Kipling 1901, 100).

Throughout the novel, Kim and the lama both search for something, the lama for his river and Kim for answer to such questions as 'Who is Kim?' (Ibid., 166, 233) and finally '[W]hat is Kim?' (Ibid., 331). These parallel searches culminate in the final pages: the lama finds his river and, in a moment of revelation, Kim finds his way in the world. But because Kim's regard for the lama is based on friendship and not discipleship, his epiphany at the end of the novel is not through Buddhism, the lama, or through any expressly religious practices or affiliation. It is essentially a secular epiphany that reinserts him into the world. How is such an experience possible? What is the legacy of this secular revelation?

There are at least two meanings to epiphany, of course. There is the Epiphany as Christian Feast commemorating both Christ's baptism and his manifestation to the gentiles. There is also the epiphany that refers to a moment of sudden intuitive understanding – epiphany as a kind of enlightenment. In the late eighteenth and nineteenth centuries, this notion of a flash of comprehension gained a larger cultural currency. In Britain, for example, Evangelism emphasized the possibility of a powerful 'transfiguring experience' – a conversion that was analogous to a kind of rebirth, a sudden transformation that gave one a new sense of purpose and direction in a life devoted to Christ. In the early nineteenth century, as M.H. Abrams (1971) has argued, Romantic writers transformed the notion of a flash of intuition and gave it more expressly secular connotations. Writers such as Wordsworth, Byron and Shelley reformulated 'religious concepts, schemes, and values which had been based on the relation of the Creator to his creature and creation . . . within the prevailing two-term system of subject and object, ego and non-ego, the human mind of consciousness and its transactions with nature' (Ibid., 13). Instead of being abandoned, religious concepts such as epiphany and revelation shifted from the sphere of organized religion to be refurbished for life in the world.

By the late nineteenth century the idea of secular revelation became an essential part of novels, poetry and secular biographies. The literary epiphany – what Edward Said (1993, 143) calls a 'regrasping of life scene' – is evident in such British novels as Charlotte Brontë's *Jane Eyre*, George Eliot's *Middlemarch* and Henry James's *A Portrait of a Lady*.[16] A moment of secular revelation also structures Kipling's own autobiography. In *Something of Myself*, Kipling recounts how his 'seven years' hard' labour in India was rewarded by a revelation of sorts.

> It happened one hot-weather evening, in '86 or thereabouts, when I felt that I had come to the edge of all endurance. As I entered the empty house in the dusk there was no more in me except *the horror of a great darkness*, that I must have been fighting for some days. Late at night I picked up a book by Walter Besant which was called *All in a Garden Fair*. It dealt with a young man who desired to write; who came to realise the possibilities of *common things seen*,

and who eventually succeeded in his desire. What its merits may be from today's standpoint I do not know. But I **do** know that *that book was my salvation in sore personal need and with the reading and re-reading it became to me a revelation, a hope and a strength.* I was certainly, I argued, as well equipped as the hero and – and – after all, there was no need for me to stay here for ever. I could go away and measure myself against the doorsills of London as soon as I had money. Therefore I would begin to save money, for I perceived that there was absolutely no reason outside myself why I should not do exactly what to me seemed good. For proof of my revelation I did, sporadically but sincerely, try to save money, and I built up in my head – always with the book to fall back upon – a dream of the future that sustained me.

(Kipling 1937, 39–40; italics mine)

This passage suggests the secular and worldly nature of Kipling's revelation as well as the important role a book – a novel – played in that epiphanic moment. A secular text offers the young man guidance – it presents a version of secular faith. The terms 'salvation' and 'revelation' have shifted from an expressly religious context and now give meaning to a life devoted to writing, to a life concerned with 'common things seen', to life in the world. And although it is fictional, the text Kipling reads seems to present something of the truth; the novel as a genre has acquired a crypto-religious status, conforming to Thomas Carlyle's dicta that it provide readers ' "edification", "healing", "guidance" and "a divine awakening voice" ' (Qualls 1982, 1). In Kipling's version of himself, the budding novelist emerges from his reading with a new direction. The novel provides not only – or not even – a structure for his life; the epiphanic moment gives a structure to Kipling's account of that life. Kipling is able to make sense out of his life through an established interpretive framework; a textual 'enlightenment' gives his (textual) self coherence.

Just as Kipling despairs and is repaired, Kim too suffers and is cured. As in Kipling's account of his own life (which was written after *Kim*), the character Kim experiences a physical version of a dark night of the soul, after which he determines 'I must get into the world again' (Kipling 1901, 331). Kim's recovery from this physical collapse precipitates a psychic confusion: he becomes 'unable to take up the size and proportion and use of things' (Ibid.). The nature of his disorientation is expressed in mechanical terms: 'his soul was out of gear with his surroundings – a cog-wheel unconnected with any machinery, just like the idle cog-wheel of a cheap Beheea sugar-crusher laid by in a corner' (Ibid.). When the solution to this difficulty arrives, when, that is, the epiphany comes, it arrives 'of a sudden', and 'with an almost audible click he [feels] the wheels of his being lock up anew on the world without' (Ibid.). With this revelation, roads, houses, cattle and people become things that are 'all real and true – solidly planted on the feet – perfectly comprehensible – clay of his clay, neither more nor less' (Ibid.). Kim's secular revelation places him securely within both the things of this world and the colonialist machinery: he will work for the Raj, he will work in the world and for what seems to be the good of India.

The nature and significance of Kim's epiphany is further emphasized by its contrast with the culmination of the lama's search. When the lama finds his river and thus the end of his search, his realization is first narrated by those who recognize only its physical manifestation. The earthy Sahiba begs the question: '[To] go roving into the fields for two nights on an empty belly – and to tumble into a brook at the end of it – call you *that* holiness?' (Ibid., 325; emphasis in original). While Kim's epiphany is dramatized in the physical and active terms valorized by the novel, the lama describes his own revelation in metaphoric and abstract language: 'Yea, my Soul went free, and wheeling like an eagle, saw indeed that there was no Teshoo Lama nor any other soul. As a drop draws to water, so my Soul drew near to the Great Soul *which is beyond all things*. . . . By this I knew the Soul had passed beyond the illusion of Time and Space *and of Things*' (Ibid., 337; emphasis mine). But as newly made man of the world and not the spirit, Kim can only understand the lama's revelation in physical terms. He finds it marvellous that the lama abstained from food for two days; told about the discovery of the river, Kim can only ask, 'Wast thou very wet?' (Ibid., 338). The divergent epiphanies emphasize what was already evident: while the lama seeks escape from the 'Wheel of Life' and its sufferings, Kim cannot keep his eyes and affections from it (Ibid., 260). From the lama's perspective, 'this a great and terrible world' (Ibid., 242). Kim, in his attachment to the lama, cannot help but revise this estimation: 'the Holy One . . . is right – [it is] a great and *wonderful* world' (Ibid., 273; emphasis mine).

Kipling's *Kim* collects – and sometimes misrepresents[17] – available knowledge on Tibetan religion; most significantly for the making of the myth of transformation in Tibet, Kipling's novel depends upon the possibility of epiphany. With that, Kipling's novel, popular in its own day, becomes part of the baggage of books carried by members of the Younghusband Expedition who travelled to Tibet in 1903 and 1904.

LHASA AT LAST

The Younghusband Expedition was largely the brainchild of Viceroy George Curzon and was designed to assert British authority in Tibet at a critical moment: Chinese power in the region had declined, Tibetan nationalism under the Thirteenth Dalai Lama was on the rise, Tibetans had encroached on what the British felt was their territory in Sikkim, and the British feared Russian influence in the region. At one point the expedition commanded by Francis Younghusband included as many as 8,000 soldiers of the Indian Army, who fought against ill-equipped Tibetan soldiers along the way. The expedition was ultimately bound for Lhasa, a place no British official had ever reached, a city that seemed to reside only in fiction. Only there, Curzon felt, could Younghusband negotiate with the Thirteenth Dalai Lama himself. The Tibetan spiritual and political leader, however, foiled these plans by fleeing to Mongolia. Determined to reach some kind of settlement, Younghusband negotiated a treaty that was ratified by the Tibetan assembly and the Dalai Lama's regent, 'Ti' Rinpoche.[18] In 1910, Younghusband published his account of the journey, *India and Tibet*.[19]

Like the writers on Tibet before and after him, Younghusband relies on a number of texts about Tibet, including those of the Japanese monk Ekai Kawaguchi, Bogle, Turner, Waddell and Kipling. From these texts, Younghusband inherits two things: a certain suspicion about Tibetan Buddhism as a living religion, and the possibility of a significant experience, the meaning of which is grounded not in organized religion but in secular relationships.

Younghusband's textual attitude towards Tibetan religion is evident in such judgments as the following: ' "The religion of the Tibetans is grotesque, and is the most degraded, not the purest form of Buddhism" ' (Seaver 1952, 247). The influence of Waddell's distrust of the Tibetan clergy is evident in Younghusband's descriptions of 'Lhasa monks'; he writes that they 'were a dirty, degraded lot, and we all of us remarked how distinctly inferior they were to the ordinary peasantry and townsmen we met' (Younghusband 1910, 266). Having studied Monier Monier-Williams' *Buddhism*[20] and the writings of Waddell, Younghusband feels confident enough to differ with Tibetan monks about the meaning of their religion. When several monks tell him that they wish to keep the Indian army out of Lhasa in order to preserve their religion, Younghusband counters, 'As the Buddhist religion nowhere preaches this seclusion, it was evident that what the monks wished to preserve was not their religion, but their priestly influence' (Ibid., 166). Relying on the work of Ekai Kawaguchi, a Japanese monk whose *Three Years in Tibet*, describing life in Sera monastery, had been published in English in 1909, Younghusband argues that 'the evil of Lamaism is that it has fostered lazy self-repose and self-suppression at the expense of useful activity and self-realization' (Younghusband 1910, 315). Indeed, 'useful activity and self-realization' – duty and epiphany – are part of Younghusband's creed. Given the judged oppressiveness and degradation of Tibetan religion, it seems clear that Tibetan Buddhism as it is lived out is not going to become part of Younghusband's faith.

However, if Younghusband is heir to suspicions about the clergy from Britain's version of Tibetan Buddhism, he also inherits from the texts of Bogle, Turner and Kipling the possibility of a significant *human* relationship between lama and Anglo. And sure enough, Younghusband is impressed by one of the lamas he meets, Ganden Tri ('Ti') Rinpoche, the Dalai Lama's regent, whom Younghusband describes as a 'benevolent, kindly old gentleman, who could not hurt a fly if he could have avoided it. No one could help liking him, but no one could say that he had the intellectual capacity we would meet in Brahmins in India, or the character and bearing one would expect in the leading men of the country. And his spiritual attainments, I gathered from a long conversation I had with him after the Treaty was signed, consisted mainly of a knowledge by rote of vast quantities of his holy books' (Ibid., 310). Younghusband's suggestion that '[n]o one could help liking him' echoes Bogle's estimation of the Teshu Lama over a century earlier: 'he is so universally beloved . . . not a man could find it in his heart to speak ill of him' (Markham 1879, 84). And like Kipling's gentle lama, Younghusband's lama has a propensity for rote quotation. A dignified, pacifist Tibetan with spiritual gifts less impressive than his humanity – we have seen this lama before. And if the resemblance is not already apparent, Younghusband goes on to assert that Ti Rinpoche 'more nearly approached Kipling's Lama in "Kim" than any other Tibetan I met' (Younghusband 1910, 325). It is through this intertextual lama that Younghusband's epiphany is set in motion.

After all the negotiations are complete, Younghusband, along with other officers, goes to bid farewell to Ti Rinpoche. Departures have been important moments between servants of the company and Tibetan lamas, and Colonel Younghusband's leave-taking is no exception. Ti Rinpoche gives the British officers a small statue of the Buddha. Younghusband reports:

> We were given to understand that the presentation by so high a Lama to those who were not Buddhists of an image of the Buddha himself was no ordinary compliment. And as the revered old Regent rose from his seat and put the present in my hand, he said with real impressiveness that he had none of the riches of the world, and could only offer me this simple image. Whenever he looked upon the image of Buddha he thought only of peace, and he hoped that whenever I looked on it I would think of Tibet. I felt like taking part in a religious ceremony as the kindly old man spoke those words; and I was glad that all political wranglings were over, and that now we could part as friends man with man.
>
> (Younghusband 1910, 325–6)

In this passage, Younghusband takes care to point out that he is not a Buddhist; in his desire for a religious ceremony, then, a Tibetan Buddhist ritual is clearly off limits. But Younghusband does have access to a kind of secularized spiritualism: he, as a reader of Kipling and a lover of Wordsworth and Byron, is heir to a discourse of the possibility of significant and transformative experience *in the world*. After leaving Ti Rinpoche, Younghusband goes out into the mountains, looks at the sky and the city of Lhasa and experiences a sense of elation.

> The scenery was in sympathy with my feelings; the unclouded sky a heavenly blue; the mountains softly merging into violet; and, as I now looked towards that mysterious purply haze in which the sacred city [Lhasa] was once more wrapped, I no longer had any cause to dread the hatred it might hide. And with all the warmth still on me of that impressive farewell message, and bathed in the insinuating influences of the dreamy autumn evening, I was insensibly infused with an almost intoxicating sense of elation and goodwill. This exhilaration of the moment grew and grew till it thrilled through me with over-powering intensity. Never again could I think evil, or ever again be at enmity with any man. All nature and all humanity were bathed in a rosy glowing radiancy; and life for the future seemed naught but buoyancy and light.
>
> Such experiences are only too rare, and they but too soon become blurred

in the actualities of daily intercourse and practical existence. Yet it is these few fleeting moments which are reality. In these only we see real life. The rest is ephemeral, the insubstantial. And that single hour on leaving Lhasa was worth all the rest of a lifetime.

(Younghusband 1910, 326–7)

In this ornate passage, the textualization of epiphany is made plain. The experience of epiphany or revelation – a structure rooted in Christian tradition, emphasized in Evangelism, shifted to secular writing, popularized in Victorian novels, and transferred from Kipling's account – makes its first appearance in an account of travel to Tibet. And it is to have important effects.

Whereas Kim's epiphany connects him to the world of things and to practical matters, Younghusband's epiphany denies the reality of the material world in favour of an inner truth realizable through contemplation of nature: 'it is these few fleeting moments which are reality.' If anything, Younghusband's vision more closely resembles that of Kipling's lama, whose realization gives him a mystical and encompassing view of the Indian empire. But with Younghusband's assertion that the things of the world are ultimately 'insubstantial', he plays down the importance of his own participation in the British imperial project, whose practices are coming to face criticism at home and in India. Indeed, in the remaining pages of *India and Tibet*, imperialism takes on mystical dimensions as Younghusband argues that 'some inward compulsion from the very core of things' drives Britons to assert their authority in places like Tibet (Ibid., 434). An 'inner necessity . . . surging up from the inmost depths of our beings' compels Britons to deal with 'weak and disorderly people' (Ibid., 435, 437). And for Younghusband, because the British feel a natural impulse towards harmony, unity and order, they cannot tolerate isolationist Tibet; they 'find they have to intervene to establish order and set up regular relations – they are, in fact, driven to establish eventual harmony, even if it may be by the use of force at the moment' (Ibid., 436, 437).

Drawing on the structure of epiphany, Younghusband finds the justification for imperialism deep within himself; his spiritual moment is put to work on behalf of the British empire. Refashioning various textual precedents, Younghusband writes a place for a coherent self, and, in effect, his 'inmost being' is colonized. Imperialist discourse clears out a space, a territory *within* Younghusband, as it were. Unlike Bogle before him, and more like Kim, Younghusband makes the world and himself cohere; by finding a place in the narrative, he finds a fit in the world. Travel becomes transformation. And the myth of epiphany in Tibet inherits a burdensome legacy.

BEYOND *INDIA AND TIBET*

In his biography *Francis Younghusband: Explorer and mystic*, George Seaver argues that the colonel kept quiet about this transformative experience for some years.[21] But after *India and Tibet* is published in 1910, Younghusband no longer remains quiet about his 'spiritual experience on the mountain-side overlooking Lhasa' (Seaver 1952, 248). As Seaver notes, 'when in after-life he was called upon to address the youth of England in schools and colleges it was this experience that he chose for his theme in preference to his more obviously exciting adventures' (Ibid.). Much of the text of Younghusband's speech from 1938 quoted by Seaver comes out of *India and Tibet*, but the mystical experience is further elaborated:

> Elation grew to exultation, and exultation to an exaltation which thrilled through me with overpowering intensity. I was beside myself with unintelligible joy. The whole world was ablaze with the same ineffable bliss that was burning within me. I felt in touch with the flaming heart of the world. . . . I was boiling over with love for the whole world. I could embrace every single human being. And henceforth life for me was naught but buoyancy and light.
>
> (Seaver 1952, 249)

In this version of the story, Younghusband notes his desire to communicate this feeling to others, to bring them to the same feeling: 'I had visions of a far greater religious faith yet to be, and of a God as much greater than our English God as a Himalayan giant is greater than an English hill' (Ibid.). Indeed, after his retirement to England in 1909, Younghusband would try to elaborate 'the intellectual framework of my conception of the universe' through continuing study of world religions, philosophy and anthropology (Ibid.). Asserting that evolution was leading humans on to greater spiritual development, Younghusband devoted himself to spiritual 'exploration'. Troping such study as a kind of travel, Younghusband writes: 'The exploring spirit was on me. I would go in front and show the ways across the spiritual unknown' (Ibid., 275). In 1919 Younghusband became the president of the Royal Geographical Society and promoted Himalayan mountaineering with a passion. Declaring that 'Man's supreme adventure in the material world was seen to be symbolical of supreme adventure in the realm of spirit', Younghusband left the mark of his spiritual-exploration ideas on later conceptions of Everest exploration (Ibid., 371).

Younghusband's fascination with Tibet did not abandon him in his final days; indeed, the image of the Buddha given to Younghusband by Ti Rinpoche had remained at his bedside throughout his final illness. As his friend, Lady Lees, reports, 'He treasured that little image more than all his earthly possessions, and as he lay in his coffin his daughter placed it on the lid' (Ibid., 374). On his tombstone was carved a relief of Lhasa with the words: 'Blessed are the pure in heart, for they shall see God' (Ibid., 375).

What did Younghusband leave for subsequent travellers to Tibet? A paradigm. A structure. The dream of transformation. For if Younghusband accomplished his spiritual epiphany at a particular moment in imperial history, the space described and inscribed in his description of the movement, once opened, would not be closed again. He 'pioneered', if you will, a certain notion of transformation in Tibet; he argued for a certain correspondence between the world and the inner life of 'man'. He opened up a territory, set a horizon of expectation through which Tibet became the place to which Westerners might travel to attain self-realization. Although Younghusband's realization comes at a particular historical and cultural moment, it becomes a structure and an expectation that moves transhistorically, or rather is ahistorically used and refashioned by later travellers. This spiritual realization becomes part of Western knowledge on Tibet; it exerts a pressure on later accounts of Tibet in English. Just as yaks, Tibetan 'sky-burial', monks and mountains become part of the baggage handled by successive travel writers, so too does inner transformation become part and parcel of subsequent travel accounts of Tibet in English. That Younghusband's realization so baldly serves the imperialist project, that his notion of himself is integrally connected to imperial discourses, is part of the legacy borne by subsequent readers and writers.

The legacy of Tibetan epiphany can be read in James Hilton's novel *Lost Horizon* (1933), Frank Capra's 1936 film version and Mark Frutkin's 1991 novel, *Invading Tibet*, as well as in such travel accounts as Peter Matthiessen's *The Snow Leopard* (1978) and Pico Iyer's *Video Night in Kathmandu* (1988), among others. It can also be seen in one recent attempt to unravel the West's fascination with Tibet, Peter Bishop's *The Myth of Shangri-La: Tibet, travel writing and the Western creation of sacred landscape* (1989). In this important study, Bishop promises to unravel the West's making of the myth of Shangri-La but ultimately serves to reinstate it. This is possible because for Bishop, as for Hodgson, Waddell and company, texts offer access to the real; travel texts matter because they offer 'a direct encounter with that elusive place' (Bishop 1989, 149). Forgoing an analysis of texts as linguistic constructions guided by convention, Bishop largely ignores the connections among texts as well as among texts and other discourses. This is necessary, of course, if Bishop is going to argue that 'Tibet seemed always to have the ability slightly to elude the total embrace of Western Orientalism. It always sustained an independent Otherness, a sense of superiority, albeit limited' (Ibid., 145). Although he is careful to cite Said's *Orientalism* in his introduction, Bishop's argument unreflectively relies on the Orientalist trope of Tibet as a special preserve. Departing from the cultural and historical critique he proffers in the opening pages, Bishop uses the depth psychology of James Hillman to imagine Tibet as an *axis mundi* around which the hopes and fantasies of the West turn; under Bishop's care, Tibet becomes an archetype which can remythify the West, give the West meaning and hope, and connect the West to its deepest self. In *The Myth of Shangri-La*, Tibet is remythified; Tibet is reinstated as secular quasi-religion. Enlightenment is once again possible.

In some ways Bishop's approach may seem to be a geographical one, for he finally eschews concerns with textuality in favour of his attention to the *place* of Tibet. I would argue that both moves should occur: one must explore the textual status of accounts just as one should attend to the specificity of Tibet as signifier. In order to explore the development of the myth, texts must be seen as something other than windows on the world or gateways to the truth; their conventional, generic, historical and linguistic status must be recognized – their connections to one another must be explored. Further, the specificity of their referents – the places which they would represent – must be considered. Without these perspectives, scholars and travellers may simply continue to cover the same territory. Without such views, we may continue to tell ourselves a story in which Tibet is merely a holy relic, the path is well-trodden, and the enlightenment we find is just a figure from another traveller's tale.

NOTES

1 Donald S. Lopez, Jr, has written thoughtfully on the special place Tibet occupies in the Western imagination. His book *Prisoners of Shangri-La: Tibetan Buddhism and the West* (1998) appeared while this book was in production.

2 Not the result of a single dramatic gesture, Tibet's inaccessibility to Europeans was a gradual process: by the mid-eighteenth century, the areas under the control of the Dalai Lama came under the suzerainty of Manchu China. China's power and thus Tibet's isolation were further consolidated by the Gurkha Wars between 1788 and 1816. Eager for trade with China, the East India Company and later the British Raj maintained a policy of non-intervention in Chinese affairs in Tibet which lasted until the 1860s. For an account of Britain's relations with Tibet in the colonial period, see Lamb (1986).

3 Lobsang Palden Yeshe, the Third Panchen Lama, had written to Hastings to ask him not to intervene in military affairs between Nepal and Bhutan. Hastings saw in this overture an opportunity to extend trade relations and sent Bogle to Tibet to reply in person to the message. Bogle was to examine trade possibilities and to seek a new land route to China.

4 British writers used various spellings of this name – Teshu, Teshoo, Tashi. This figure is better known these days as the Panchen Lama, whose monastery was Tashilhunpo. He is considered second only to the Dalai Lama, and a manifestation of Opame, or Amitabha, the future Buddha.

5 A curious feature of this borrowing is that Turner notes that Bogle found the Teshoo Lama 'open, candid, and generous in the extreme' (338) and attributes the same qualities to the young incarnation.

6 Turner's account of his travels to Xigatse, published in 1800, would remain the only first-hand account of Tibet in English until the mid-nineteenth century. Bogle's account was only published in 1876 by Clements R. Markham, who collected several accounts of Tibet during a period of keen British interest in Tibet. Included in this volume were the unedited journals of Thomas Manning, who travelled to Lhasa in 1811–12. A Sinophile known for his eccentric manner, Manning longed to reach Beijing and travelled without Company sanction.

7 I adapt this phrase from Wendy Doniger O'Flaherty's *Other People's Myths*, New York: Macmillan, 1988.

8 Hodgson made a name for himself when he requested and received from the Fourth Panchen Lama a complete set of the Tibetan religious corpus known as the *Kagyur* and *Tengyur*, Tibetan translations of Sanskrit Buddhist texts and commentaries on them in Tibetan.

9 Sakyamuni Buddha probably spoke the vernacular Magadhi, and the earliest Buddhist texts were written several centuries after the Buddha's death in both Sanskrit and Pali.

10 Alexander Csoma de Körös had made his way from Transylvania to Central Asia in the 1820s in search of the origins of the Magyar people. Encouraged to study Tibetan by Thomas Moorcroft, another Himalayan adventurer, Csoma de Körös

lived and worked in a Tibetan Buddhist monastery in Zanskar, where he compiled a Tibetan–English dictionary, a Tibetan grammar and accounts of Tibetan literature and history. He eventually accepted a small stipend from the East India Company for his efforts and in 1837 became a librarian at the Asiatic Society in Calcutta. See Duka (1885).

11 Such an attitude operated in places other than Tibet and is apparent in the work of the Theosophists Helena Blavatsky and Henry Steele Olcott in Ceylon (now Sri Lanka). Relying on their textual researches in Buddhism, Blavatsky and Olcott found Ceylonese Buddhism degenerate and worked to sort out 'real' Buddhism from its cultural and historical accretions.

12 Waddell writes, '[R]ealizing the rigid secrecy maintained by Lamas in regard to their seemingly chaotic rites and symbolism, I felt compelled to purchase a Lamaist temple with its fittings; and prevailed on the officiating priests to explain to me in full detail the symbolism and the rites as they proceeded' (1895, viii).

13 Throughout the latter half of the nineteenth century in Britain and the United States, similar claims were made about Christianity's devolution from the time of Christ.

14 Philip Almond describes Europe's version of Buddhism in detail. Given a definition of religion which insisted on the necessity of a god, Europeans also argued over whether Buddhism might not be better termed a philosophy and not a religion at all (1988, 94).

15 For more on European conceptions of Tibetan religion, see Donald S. Lopez (1995, 1996).

16 Barry V. Qualls's *The Secular Pilgrims of Victorian Fiction* (1982) explores how pilgrimage and conversion are reshaped for secular use in the Victorian novels of Carlyle, Dickens, Charlotte Brontë and George Eliot.

17 For example, Kipling repeatedly has the lama utter proverbs in Chinese (92, 93, 119, 281); at one point the lama is said to have 'slid into Tibetan and long-droned texts from a Chinese book of the Buddha's life' (80). Chinese and Tibetan Buddhist traditions have quite different histories and literatures. Kipling also clothes the lama in yellow robes (119). Tibetan monks, unlike most Buddhist monks in the world, have long worn maroon robes.

18 Initially, the treaty provided for new trade marts, imposed a large indemnity on the Tibetans, allowed for British occupation of the Chumbi Valley until the indemnity was paid, and forbade the Tibetans from having any dealings with any foreign power without British consent. The terms and the severity of this treaty would be questioned by authorities in Britain who felt that Younghusband had acted without proper advisement; indeed, Younghusband was censured for taking initiative in determining the terms of the treaty. The binding status of this treaty was also questioned by British, Tibetan and Chinese officials. Eventually, the original treaty was abandoned for a much more lenient one – one which Younghusband felt was ineffective.

19 Other accounts of the expedition include L. Austine Waddell's *Lhasa and Its Mysteries* (1905) and Edmund Candler's *The Unveiling of Lhasa* (1905).

20 Younghusband's biographer, George Seaver, notes that Younghusband had taken this volume with him on an expedition to Kashmir along with 'Lubbock's *Ancient Civilization*; some novels of Dickens; Momerie's *Sermons*; a Bible and Prayer Book; besides various official reports, and the R.G.S.'s invaluable *Hints to Travellers*' (Seaver 1952, 113).

21 Also see Patrick French's biography of Younghusband (1994).

REFERENCES

Abrams, M.H. (1971) *Natural Supernaturalism: Tradition and revolution in Romantic literature*, New York: Norton.

Almond, Philip (1988) *The British Discovery of Buddhism*, Cambridge: Cambridge University Press.

Bishop, Peter (1989) *The Myth of Shangri-La: Tibet, travel writing and the Western creation of sacred landscape*, Berkeley: University of California Press.

Candler, Edmund (1905)*The Unveiling of Lhasa*, London: Edward Arnold.

Capra, Frank (1936) *Lost Horizon* (film).

Duka, Theodore (1885 [1971]) *Life and Works of Alexander Csomo de Körös*, New Delhi: Manjusri.

Frutkin, Mark (1991) *Invading Tibet*, New York: Soho Press.

Hilton, James (1933) *Lost Horizon*, New York: Grosset and Dunlap.

Hodgson, Brian (1972) *Essays on the Languages, Literatures, and Religion of Nepal and Tibet. Together with Further Papers on the Geography, Ethnology, and Commerce of those Countries*, New Delhi: Manjusri.

Hovell, Laurie (1993) 'Horizons Lost and Found: Travel, writing, and Tibet in the Age of Imperialism', Ph.D. thesis, Syracuse University.

Iyer, Pico (1988) *Video Night in Kathmandu*, New York: Vintage.

Kawaguchi, Ekai (1909) *Three Years in Tibet*, Adyar: Theosophical Society.

Kipling, Rudyard (1901 [1989]) *Kim*. New York: Penguin.

—— (1937 [1990]) *Something of Myself*, Cambridge: Cambridge University Press.

Lamb, Alistair (1986) *British India and Tibet*, New York: Routledge and Kegan Paul.

Lopez, Donald S., Jr (1995) 'Foreigner at the Lama's feet', in *Curators of the Budd'ha*, Chicago: University of Chicago Press.

—— (1996) 'Lamaism and the disappearance of Tibet', *Society for the Comparative Study of Society and History* 38, pp. 3–25.

—— (1998) *Prisoners of Shangri-La: Tibetan Buddhism and the West*, Chicago: University of Chicago Press.

Markham, Clements (1879 [1971]) *Narratives of the Mission of George Bogle to Tibet and of the Journey of Thomas Manning to Lhasa*, New Delhi: Manjusri.

Matthiessen, Peter (1978) *The Snow Leopard*. New York: Penguin.

Qualls, Barry V. (1982) *The Secular Pilgrims of Victorian Britain: The novel as book of life*, Cambridge: Cambridge University Press.

Said, Edward W. (1979) *Orientalism*, New York: Vintage,

—— (1993) *Culture and Imperialism*, New York: Knopf.

Seaver, George (1952) *Francis Younghusband: Explorer and mystic*, London: John Murray.

Turner, Samuel (1800 [1981]) *An Account of An Embassy to the Court of the Teshoo Lama in Tibet*, New Delhi: Manjusri.

Waddell, L. Austine (1895 [1979]) *Buddhism and Lamaism of Tibet*, New Delhi: Heritage.

—— (1905 [1988])*Lhasa and Its Mysteries*, New York: Dover.

Younghusband, Francis (1910 [1971]) *India and Tibet*, Delhi: Oriental Publishers.

4 Writing Travel and Mapping Sexuality

Richard Burton's Sotadic Zone

Richard Phillips

> After much wandering, we are almost tempted to believe the bad doctrine that morality is a matter of geography.
>
> (Burton 1860, 84)

Victorian travel writers charted imaginative terrain on which new sexualities – including homosexuality and heterosexuality – were constructed, and also contested. Few travel writers took a greater interest in sex than did Sir Richard Francis Burton (1821–90), and none played a greater part in mapping and contesting sexualities. A popular travel writer and translator as well as a famous geographer, who identified himself in print as a Fellow then Gold Medalist of the Royal Geographical Society, Burton was also acknowledged as a pioneer sexologist. His travel, translation, geography and sexology were intimately related. In the geography of his travels and translations, he mapped sexualities. In particular, he mapped a 'Sotadic Zone' in which, he claimed, pederasty was common. There, in his most explicitly sexual geography, Burton was able to conceptualize a form of male homosexuality and chart a relationship between this marginalized homosexuality and the dominant, heterosexual sexuality of the material and metaphorical centre – England. I will argue that Burton used travel, and specifically travel geography, as a medium in which to contest contemporary constructions of sexuality, and more specifically to protest against contemporary homophobia.

The geographers who are currently pushing sex and sexuality onto the discipline's agenda, while exploring constructions of sexuality and resisting homophobia, may find in Burton something of an intellectual ancestor. David Bell's guest editorial in *Society and Space*, '[screw]ING GEOGRAPHY' (Bell 1995),[1] is reminiscent of much that Burton wrote. Like Burton, Bell writes as a geographer and addresses geographers (and others). Both write about sex, openly and broadly. Like Burton, Bell assumes a combative style and complains of censorship – making much of the claim that '[screw]ING' should have read 'FUCKING' in the Association of American Geographers conference programme. Like Burton, Bell is aware that his outspoken and unorthodox, anti-establishment approach may threaten his career and livelihood. Like Burton, he declares his readiness to eschew and provoke the contemporary establishment, although few could hope to provoke the establishment as comprehensively as Burton, who made himself *persona non grata* in halls of power from the Foreign Office to the Royal Geographical Society. Indeed, few could hope, or want, to be as provocative as Burton often was. Geographers of sex and sexuality can only be ambivalent about their relationship to a figure whose sexual politics were mixed up with anti-Semitic and anti-African racism, Francophobia, misogyny, violence (not all consensual) and enthusiastic imperialism. Burton's sexual geography does, however, raise and explore a number of important questions about relationships between geography and sexuality: the somewhat parochial question of why geographers should concern themselves with sexuality; where sex belongs on the geographical agenda; and more importantly, the question of what difference space makes to sex and sexualities. Burton raises and answers some questions about how sexualities are constructed and contested, how they are mapped.

Mapping, especially when metaphorical, is deceptively slippery.[2] Maps seem to be scientific, objective, impersonal, even mechanical images. They assert what Svetlana Alpers calls an 'aura of knowledge' (1983, 133). 'They depict', as Neil Smith puts it, 'the world as it "really is"; as the author vanishes in the map, the map exudes authority' (1994, 499). Maps, naturalized as facts, are received with trust. Those who actively and self-consciously 'read between the lines of the map', along the lines advocated by map critic J.B. Harley (1992, 233), are probably an intellectual and/or critical minority. Maps construct taken-for-granted worlds in which geographies and identities are naturalized. They seem to provide firm ground upon which to stand, a sense of security for those who like to know where they are. But maps are more ambivalent than this, more open. Like other texts, their meanings are neither fixed nor singular. They can be slippery, as Deleuze and Guattari propose:

> The map is open and connectable in all of its dimensions; it is detachable, reversible, susceptible to constant modification. It can be torn, reversed, adapted to any kind of mounting, reworked by individual group or social formation.
>
> (Deleuze and Guattari 1988, 12)

A map such as this is enabling rather than confining, a point of departure rather than an end in itself. Burton's sexual map is a case in point. On first inspection, it is an orthodox map – authoritative and objective, factual and static, bounded and divided – anything but a space of travel. However, closer inspection reveals its debt to Burton's travelling perspective and presents the geography as more a point of departure, a starting point for explorations of sexuality, in which all is fluid and boundaries are set up only to be crossed.

Burton's most sustained sexual discourse, in which he mapped the Sotadic Zone, came in the translated stories, footnotes and 'Terminal Essay' in his *Plain and Literal Translation of the Arabian Nights' Entertainments or The Book of a Thousand Nights and a Night* (1885–6).[3] Published in his old age, the *Nights* confirmed Burton's reputation as a translator, although he had already produced a number of translations, including the *Kama Sutra of Vatsyayana* (1883) and the *Ananga-Ranga or the Hindu Art of Love* (1885). Earlier in his life, Burton was known mainly as a travel writer and explorer, the author of *Personal Narrative of a Pilgrimage to El-Medinah and Meccah* (1885–6), *First Footsteps in East Africa; or, An Exploration of Harar* (1856) and *The Lake Regions of Central Africa, A Picture of Exploration* (1860).[4] Burton's credibility as a geographer suffered after his expedition in search of the source of the Nile, when he entered into a public and undignified dispute with his travelling companion, John Hanning Speke, a dispute which culminated in the apparent suicide of the latter, and which led Burton to make some polemically combative, overstated and implausible claims about African geography.[5] His career as an explorer and scientific geographer, who enjoyed the occasional support of the Royal Geographical Society, took a blow. Thus began a period when Burton scribbled, dabbled, drank and passed the time in a series of diplomatic outposts, where he lived in virtual exile, alone or with his wife Isabel. It was only after a heart attack, late in life, that he re-established himself, this time as a translator. Throughout his life Burton remained a geographer and an expert on erotica, although the balance between these shifted from the former to the latter between his early travels and his late translations. In his travel narratives, Burton comes across as an erudite geographer. In his translations, he is more a geographical sexologist. In his travel narratives, he allowed himself to be censored by publishers,[6] and he relegated sex to the margins of his work. In his translations, he allowed himself some freedom with geography and brought sex to the forefront of his work, which he printed without the interference of editors or publishing companies.

The geography of Burton's translations, especially the *Nights*, is primarily metaphorical rather than material. Although Burton claimed to describe the 'true East' (Burton 1885, 1.xiii), and justified this work to would-be censors and critics as straightforward description of the Orient, essential background reading for would-be British conquerors and administrators, the East and the Orient were primarily vehicles for his sexual imagination. The Sotadic Zone was Burton's attempt to tell 'the truth about sex',[7] and perhaps about his own sexuality, but not the 'truth' about the East – despite his insistence to the contrary. A material space of concrete sexual encounters[8] did, of course, exist, but like Burton I will focus on its imagined, metaphorical counterpart. I acknowledge, however, that as Edward Said, Rana Kabbani and others have shown, the material and metaphorical are always connected. This connection has implications for the colonized Sotadic Zone, as it does for the sexualities that are imagined there.

MAPPING SEXUALITY: THE *NIGHTS* AND THE SOTADIC ZONE

Burton's *Nights* can be read as a map of sexuality. Despite his unorthodox subject material, Burton makes what looks like an orthodox map. There are regions – neatly mapped zones with precise boundaries. These boundaries polarize geography and sexuality, separating Occident and Orient, heterosexuality and homosexuality. The author, a self-effacing translator and scholar, distances himself from the text, making it seem objective, mechanical and believable. The effect is to contain, distance and marginalize sexualities associated with the Orient.

Translating the *Nights*, Burton sexualized a collection of stories that was already a classic and a favourite in England and France (Caracciolo 1988). As he put it, he offered readers an 'uncastrated' (Burton 1885, 1.ix) version of stories that had suffered in the hands of censorious scholars, translators, editors and publishers. Burton claimed to translate without prejudice, resisting the temptation – that had plagued many of his predecessors – to sanitize and censor. In addition, he took the opportunity to add sexual and other details, in extensive if often loosely tangential footnotes. And he appended extended essays on 'Pornography' and 'Pederasty' in the tenth of what grew to sixteen volumes.[9] Burton arranged for the book to be privately printed, so he did not suffer the editorial intrusion and/or censorship of publishers, and he evaded the repercussions that might have come from formally publishing (rather than merely printing) erotica.[10] Burton's translation was more than competent, but, partly because it came in the wake of another translation, the well-crafted and literary work of John Payne,[11] it was noted primarily for its sexual content.

The sexual geography of Burton's *Nights* is spelled out most concisely in the Terminal Essay on 'Pederasty'. Pederasty has recently been defined as 'sexual activity with pubertal boys' (Hyam 1990, 226).[12] In Burton's time, however, pederasty was used more widely, to refer to relationships (not just sex) between older and younger men. Reading Burton on pederasty, one must keep the nineteenth- rather than the twentieth-century definition in mind. Although fifty discursive pages are devoted to the subject, which Burton argues is 'geographical and climatic, not racial' (1885, 10.207), the main points are summarized neatly, as follows.

1. There exists what I shall call a 'Sotadic Zone', bounded westwards by the northern shores of the Mediterranean (N. Lat. 43°) and by the southern (N. Lat. 30°). Thus the depth would be 780 to 800 miles including meridional France, the Iberian Peninsula, Italy and Greece, with the coast-regions of Africa from Marocco to Egypt.
2. Running eastward the Sotadic Zone narrows, embracing Asia Minor, Mesopotamia and Chaldæa, Afghanistan, Sind, the Punjab and Kashmir.
3. In Indo-China, the belt begins to broaden, enfolding China, Japan and Turkistan.
4. It then embraces the South Sea Islands and the New World where, at the time of its discovery, Sotadic love was, with some exceptions, an established racial institution.
5. Within the zone the Vice is popular and endemic, held at worst to be a mere peccadillo, whilst the races to the North and South of the limits here defined practice it only sporadically amid the opprobrium of their fellows

who, as a rule, are physically incapable of performing the operation and look upon it with the liveliest disgust.

(Burton 1885, 10.206–7)

Burton continues, speculating on 'physical causes' of pederasty. He suggests that 'within the Sotadic Zone there is a blending of the masculine and feminine temperaments' (Ibid., 208), and an 'abnormal distribution and condition of the nerves' (Ibid., 209). Early sexologists, attempting to develop theories of homosexuality, were to sift through Burton's writing for something useful. But, as Havelock Ellis and J.A. Symonds explained, they found Burton's theoretical sexology shallow and poorly conceived.[13] They pointed to inconsistencies and inadequacies, especially in Burton's simplistic conception of desire – as something universal, which wells up from somewhere inside us all. But if the new sexologists were nonplussed by Burton, this was partly because they were looking for a general theory of homosexuality, whereas Burton was content to write a contextual, historical geography of sexualities. Explaining that the origins of pederasty were contextal, 'lost in the night of ages' (Ibid., 210), he devoted the bulk of his essay to exploring those origins, which together form the detail – the texture and the real substance – in his map of the Sotadic Zone.

The Sotadic Zone is precisely bounded, with imaginary walls that seem to keep pederasty in, to contain this and other deviant but just-repressible desires and sexual acts. The boundaries are precisely mapped, to particular latitudes and national borders. They hold back desires that seem to well up within the zone. Like the walls of a zoo, they contain all sorts of animal passions, which are fascinating but dangerous. These include pederasty, which Burton refers to as 'Le Vice contre nature', or 'Le Vice', and its female counterpart (but not equivalent), 'Tribadism'.[14] There is 'debauchery' and 'temptation', 'erotic perversion' (Ibid., 222) and 'evil' (Ibid., 233). The list of 'abominations' (Ibid., 224) and 'corruptions' (Ibid., 233) is long. Pederasty is practised alongside infanticide and cannibalism, prostitution and bestiality (Ibid., 240). Burton is at times titillated, at times disgusted, and usually pejorative. Here, it seems, is evidence to support Isabel Burton's claim that (to paraphrase) her husband shared and endorsed his countrymen's homophobia and public sexual conformism, and that he broke discursive taboos only to condemn and to bring deviants before the eyes of the law and/or the doctor (Brodie 1967, 329). There is, as I shall explain, evidence to support wholly different readings of Burton. The point remains, though, that it was and is possible to read Burton as a conventionally homophobic, also a conventionally imperialist writer. Burton seemed to reproduce an old Orientalist fantasy – that of a sexually exciting, dangerous and threatening Orient – and fix it in real time and space.[15]

Within the Sotadic Zone, an area in the eastern Mediterranean from Greece to Egypt forms a centre of gravity, an historical centre, from which desire spreads out, threatening surrounding areas. Like an aggressive and dominant lover, the zone 'embraces' Asia and the South Pacific, 'enfolding' nations in its arms, and threatening Europe with its dangerous passions. Burton traces the origins and spread of particular forms of pederasty, from Greece to Rome, and from Rome to North Africa, for example, in each case presenting pederasty as something forced by strong, conquering nations upon their weaker counterparts (Burton 1885, 10.218–22). Pederasty is a constant threat, against which nations must defend themselves. England, in particular, must defend itself from threats of pederasty, which may come from its cities and its borders; Burton mentions 'sporadic' out-breaks of pederasty in London and endemic pederasty among ancient Celts (Ibid., 222, 246). Despite this dynamic, pederasty seems to be contained by the boundaries on Burton's map. Its geographical containment seems to define it as a discrete, containable form of sexuality.

The Sotadic Zone is distanced from England, and is both geographically and sexually disconnected. Burton employs a variety of distancing devices. Most tan-gibly, he reproduces geographical and imaginative distance between contempor-ary constructions of Occident and Orient, by pinning his Sotadic Zone roughly on the latter. He tells the reader that the geography of the *Nights*, and the Sotadic Zone more generally, is a region he imagines from afar. He says he dreamed of it while in the 'luxuriant and deadly deserts of West Africa' and the 'dull and dreary half-clearings of Brazil', where he passed years of 'official banishment' engaged in minor diplomatic duties, and where he found solace in escapist Oriental day-dreams and translations:

> From my dull and commonplace and 'respectable' surroundings, the Jinn bore me at once to the land of my predilection, Arabia, a region so familiar to my mind that even at first sight, it seemed a reminiscence of some by-gone metempsychic life in the distant past.
>
> (Burton 1885, 1.vii)

Burton further distances himself from the Sotadic Zone, and from stories set there, by assuming the seemingly passive role of the translator, who merely retells stories. In the 'Terminal Essay', Burton distances himself from language, details and anecdotes by emphatically quoting others, effectively disowning much of his own text. Pederasty, for example, is 'Le Vice' – someone else's term, which Burton uses casually, even ironically. Thus Burton puts ambiguous distance between himself and pederasty.

Burton further distances the Sotadic Zone through a series of visual images and metaphors. His cartographic frame, which places the Sotadic Zone on an imaginary map of the world, provides the first visual abstraction, while giving it a factual appearance. In other visual metaphors, Burton claims to display 'a *landscape* of magnificent *prospects* whose *vistas* are adorned with every charm of nature and art' (Burton 1885, 10.254; my emphases). He sees from what seems to be a fixed, elevated position. His vision is panoramic, a grand sweep through geography and history. He 'glanc[es] over the myriad pictures of this panorama', with a gaze that is both self-effacing and voyeuristic. Even scenes of rape are regarded with casual, light-hearted detachment. In the 'Terminal Essay', for example,

A favourite Persian punishment for strangers caught in the Harem or Gymnæceum is to strip and throw them and expose them to the embraces of the grooms and the negro slaves. I once asked a Shirazi how penetration was possible if the patient resisted with all the force of the sphincter muscle: he smiled and said, 'Ah, we Persians know a trick to get over that; we apply a sharpened tent-peg to the crupper-bone (os coccygis) and knock till he opens'.

(Burton 1885, 10.235)

Visual distance is also produced by Burton's graphic imagery. He echoes the 'Eastern story-teller' who, he claims, sees and tells all, ushering the listener into every situation, even 'the bridal chamber', where he describes 'everything he *sees and hears*' (Burton 1885, 1.xvi; my emphases). The cumulative effect is to display sex, sexuality and sexual geography that seem both realistic and remote.

The authoritative, realistic and sexually frank tone of Burton's essay and notes is licensed and augmented by the scholarly appearance he cultivates. Graphic sexual images are justified on the basis of clinical and geographical interest. Occupants of the Sotadic Zone are presented, like visitors to the doctor's surgery, in 'decent nudity'. Burton, who assumed the disguise of a doctor on some of his Oriental travels, also posed as a kind of doctor – a sexologist – in Britain. As a doctor, he made sure of getting a good look at lots of 'decent nudity', and enjoyed the medical practitioner's licence to write about it. He also enjoyed the professional geographer's licence for sexual discourse, which he exploited and extended, both in the scholarly associations and journals he co-founded, to provide outlets for 'learned debauchery' and sexually explicit ethnology,[16] and also in the geographical and anthropological footnotes and essays he attached to the *Nights*.[17] As a scholar, Burton claimed to write for 'students', and insisted that 'nothing could be more repugnant than the idea of [the *Nights*] being placed in any other hands than the class for whose especial use it has been prepared'.[18] He therefore asserted his licence to talk about sex professionally, with a freedom that would have been inconceivable in 'popular books' (Burton 1885, 1.xviii), and took the opportunity to chart something resembling a formal, scholarly map.

Thus, with self-effacing distance and scholarly licence, Burton charts a region that seems to have little to do with him, a region that is, nevertheless, a reflection of his colonial desires. In other words, he makes a typical colonial map. His role, as translator and scholar, is less passive than it appears. He selects stories, seeks out information and then decides how to retell and report, translate and annotate.[19] And, despite his stated intention of restoring the *Nights* to their original form, Burton proceeds to censor them. As Boone shows, whenever Burton 'must envision himself as a reader', and therefore become part of the picture he paints, 'he finds himself restoring the "fig-leaf" beneath which he has previously declared himself willing to glimpse' (Boone 1995, 93). Detaching himself from the picture in this way, Burton evades the question of whose desire and whose sexuality are being mapped. He implies that sexual desire is universal; men are always looking for sex, however and wherever it is available to them. To Burton, empire just seems to expand the range of choices, providing many new sexual outlets. This reading of empire and sexuality is largely echoed by Ronald Hyam (1990), the most comprehensive modern chronicler of sex in the British empire. However, Burton's conceptions of desire and sexuality are socially, geographically and historically specific. The self-effacing observer, listener and student in Burton's *Nights* is, as the author points out, 'a small section' of the British public. Unlike the oral tradition, from which Burton worked, the book was attributed to a male author and addressed to men. Women were advised not to read the book, and even Isabel – editor of an expurgated Household Edition – claimed never to have read the complete original.[20] Burton's readership was further restricted to those who could afford the high cost of subscribing (one guinea for each of the ten volumes).[21] Like the author, the typical reader of the *Nights* was a middle- to upper-class Englishman. His perspective on the Sotadic Zone was both male and colonial, and his power was reflected in what he saw there. Sex, never reciprocal, was constructed from the perspective of the powerful, dominant man. Hence Burton's interest in pederasty, in which one partner is dominant, rather than homosexuality, which may be more reciprocal.[22] Hence Burton's tendency to speak of 'the use of boys' (e.g. Burton 1885, 1.211) in place of women, in some cultures. Hence his realized and desired power over those 'boys'. Despite Burton's assumption that desire precedes power, colonial power relations are embedded in the sex he describes and the sexual geography he maps.

In the *Arabian Nights*, then, Burton appears to construct an orthodox, colonial map of a somewhat unorthodox topic. He boldly goes where few have gone before, broaching the *terra incognita* of figuratively and metaphorically marginal sex. But then, it seems, he condemns what he finds. In the geographically marginal Sotadic Zone, he marginalizes pederasty, containing it within the boundaries of a divided and static geography. 'Appears', I stress, because the rigid surfaces of Burton's geography give way, on closer inspection, to reveal a less orthodox map, one which is open and fluid, like the maps envisaged by Deleuze and Guattari, whom I have quoted above. Clues to this lie in Burton's identity as a traveller, and his career as a travel writer.

TRAVEL GEOGRAPHIES AND TRAVELLING SEXUALITIES

The rigid surfaces, static topography and immutable boundaries of Burton's Sotadic Zone belie, but lead the reader into, a more open and fluid sexual geography. Ostensibly firm and familiar ground, they construct tangible points of departure for deceptively slippery, comparatively sophisticated constructions of sexuality. Slippery and fluid, they are indebted to Burton's travels. They reflect the identity – including the infamously ambivalent sexuality[23] – of the traveller, and the style and geography of his travel narratives.

Despite the apparent separation between his career as a travel writer and that as a translator, in his youth and old age respectively, Burton's travels and translations were always connected. The *Nights* gave Burton a point of entry to the Orient and a way of seeing the land and its peoples. He tells stories from the *Nights* to travelling companions, gathered around campfires at night. In so doing, he identifies sexual and moral issues, which he returns to later in life. In *First Footsteps in East Africa*:

> When Arabs are present, I usually read out a tale from 'The Thousand and One Nights', that wonderful work, so often translated, so much turned over, and so little understood at home. The most familiar of books in England, next to The Bible, it is one of the least known, the reason being that about one-fifth is utterly unfit for translation; and the most sanguine Orientalist would not dare to render literally more than three quarters of the remainder.
>
> (Burton 1856, 26)

Again, in the preface to the *Nights*, he recalls how he assumed the role of story-teller, when travelling with Arabs.

> The Shaykhs and 'white-beards' of the tribe gravely take their places, sitting with outspread skirts like hillocks on the plain, as the Arabs say, around the camp-fire, whilst I reward their hospitality and secure its continuance by reading or reciting a few pages of their favourite tales. The women and children stand motionless as silhouettes outside the ring; and all are breathless with attention; they seem to drink in the words with eyes and mouths as well as with ears.
>
> (Burton 1885, 1.viii)[24]

Less tangibly, the *Nights* provided Burton with a way of seeing the Arab world, when he was travelling there. James and Nancy Duncan have shown how texts provide ways of *reading* the landscape (Duncan and Duncan 1988). The *Nights*, perhaps more than any other Oriental or Orientalist text, provided Burton and other Western travellers with a way of reading the landscapes of Arabia, landscapes that might otherwise have been illegible to them. A recent travel essay on Cairo, by Jan Morris, provides an unusually explicit illustration of how European travellers, steeped in the *Nights*, have read Oriental landscapes. In the article, which appeared in the British *Guardian* newspaper, Morris evokes a street near the 'great bazaar quarter' of Cairo, a street she interprets as 'the true locale of the Thousand and One Nights – ostensibly set in Baghdad but really a reflection of this tremendous oriental capital' (1996, 29). Similarly, if not always so explicitly, the people and places in Burton's narratives often seem to have stepped out of the *Nights*. Fatma Moussa-Mahmoud traces some of the characters in European Oriental travel literature to the *Nights*. For example, a travelling companion in Burton's *Personal Narrative of a Pilgrimage*, the plump unmarried twenty-eight-year-old Omar Effendi, is compared to Kamer al-Zaman of the *Nights*, while two cookmaids in *First Footsteps* are nicknamed Sheherazade and Deenarzade. The *Nights* – 'part of the furniture of his mind' – provided Burton with ready-made *dramatis personae*, who served as travelling companions and as templates for Burton's own disguises (Mousssa-Mahmoud 1988, 105).

While the *Nights* shaped the geography of Burton's travels, it is also true that Burton's travels shaped the geography of his *Nights*. By his own admission, in the preface (quoted above), the narrator of Burton's *Nights* was the traveller telling stories by the campfire, and not the detached armchair dreamer – the exiled minor diplomat or the old man in his study. Burton emphasizes this point, by claiming to have conceived and begun the translation in Aden, where (he says) he spent the winter of 1852 with a travelling companion, Dr Steinhauser (Burton 1885, 1.ix). He remembers how the *Nights* were a comfort when he was on the road, and even in the midst of adventures. 'Throughout a difficult and dangerous march across the murderous Somali country', for example, 'The *Nights* rendered [him] the best of service' (Burton 1886, 6.388). In the footnotes and essays that accompany the *Nights*, Burton reiterates and develops lines of thought that first appear in his travel narratives, as the thoughts of a traveller. In *First Footsteps*, he introduced what was to be his thesis in the *Nights*, reflecting only 'After much wandering' that 'morality is a matter of geography' (1860, 84). In the *Nights*, Burton refers frequently to his travels and first-hand observations, including those of boy brothels in Karachi, and he directs readers to his published travel books.[25] He attributes his command of Arabic and his fascination with the Orient, especially its deserts, to a 'succession of journeys and long visits' and 'an exploration' (1886, 6.416). Referring to his travels, Burton establishes his credentials as a *bona fide* Orientalist and translator, backs up some of his specific claims and explains away his interest in pederasty. He also puts his travels, and himself as a traveller, into the *Nights*, and therefore links the geography of the *Nights* to the geography of his travels. Publishing the work under his own name – the name of a famous traveller – at the risk of prosecution for obscenity, Burton effectively invites readers to make a connection between the geography of his *Nights* and the geography of his travels.

Burton's travel narratives, unlike his translations and essays, are adventure stories – heroic, exciting quests in *terra incognita*. They are introduced and in some cases subtitled as adventures, typically as records of 'personal adventure' (Burton 1860, 1.vii). They contain the elements of quest narratives, identified by Northrop Frye as 'the *agon* or conflict, the *pathos* or death-struggle, and the *anagnorisis* or discovery, the recognition of the hero, who has clearly proved himself' (Frye 1990, 187). Burton is the hero and narrator. Unlike the self-effacing translator or scholar behind the *Nights*, he is at the centre of this narrative. In *The Lake Regions*, for example, Burton explains that he has 'not attempted to avoid intruding matters of a private and personal nature upon the reader; it would have been impossible to avoid egotism in a purely egotistical narrative' (1860, 1.viii). Burton communicates the pleasure, the excitement and the danger of his adventures. He is engaged, part of the picture he describes. Adopting disguises, including the clothes of a Persian wanderer, the skin of an Oriental (darkened with walnut juice) and the languages and dialects to match, he blends into the Orient he constructs. This is not the mapped, circumscribed space of formal geography, but the theatrical *mise en scène* of a dramatic adventure. Fluid, like the landscapes of exploration that Paul Carter explores in *The Road to Botany Bay* (1988), the geography of Burton's travel consists not of absolute, fixed places, but points along a road. Burton describes the view from the road, the vistas before and behind him, and maps path-like, linear, fluid geography. Here, boundaries are to be crossed, regions to be passed through – as the name of one district, which Burton translates as 'put down! (*scil.* your pack)' (1860, 313), makes clear. This space of travel and adventure seems very different from the geography of the *Nights*, although the two are closely connected, and similar.

The geography of the *Nights*, not the contained region-in-itself it seems to be, is the setting of an Englishman's adventure. Burton identifies the *Nights* as a classic of English adventure literature, suggesting that 'Sheherazade [was] as familiar to the home reader as Prospero, Robinson Crusoe, Lemuel Gulliver and Dr Primrose' (1885, 10.95). As the setting of an adventure, Burton's Arabia is a magical place, visited for a while, or perhaps just imagined. Burton sets the scene, in the first instance, with a leap of imagination.

> Again I stood under the diaphanous skies, in air glorious as aether, whose every breath raises men's spirits like sparkling wine. Once more I saw the evening star hanging like a solitaire from the pure front of the western firmament; and the after-glow transfiguring and transforming, as by magic, the homely and rugged features of the scene into a fairy-land lit with a light which never shines on other soils or seas.
>
> (Burton 1885, 1.vii)

Hardly a civil servant's guide to the Orient, this is the setting of an adventure, reminiscent of what Joseph Campbell calls the 'zone unknown' of adventure myths:

> This fateful region of both treasure and danger may be variously represented: as a distant land, a forest, a kingdom underground, beneath the waves, or above the sky, a secret island, lofty mountaintop, or profound dream state; but it is always a place of strangely fluid and polymorphous beings, unimaginable torments, superhuman deeds, and impossible delight.
>
> (Campbell 1949, 58)

A conventionally heroic explorer and map-maker, Burton mapped the Sotadic Zone in what, he claimed, was virtual *terra incognita*. In keeping with his image as Britain's most Nietzschean adventurer, who followed the philosopher's instructions to 'live dangerously!' and 'send your ships into uncharted seas' (see Zweig 1974, 204), Burton liked to be seen to go to new areas, where others feared to tread. Earlier in his life he had travelled to geographical *terrae incognitae*, parts of Africa and Arabia that remained as 'huge white blots' on the map (1856, 1.1). In his old age, he travelled to metaphorical *terrae incognitae*, confronting silences and blanks of another sort – silences of sexual discourse. Silences there were, despite the proliferation of sexual discourse, which Michel Foucault has identified, and despite pervasive, popular erotic myths of Orient. Sexual discourse was restricted largely to the legal and medical professions, to religion, 'serious' literature and underground pornography.[26] When doctors, prosecutors, religious leaders and purity campaigners[27] opened up sexual discourse in the 1880s, they did so in order to specify what 'Thou Shalt Not' do. They brought homosexuality from obscurity into the glare of the public eye, and brought homosexuals into the law courts and prisons.[28] When Burton broke his long, if not total silence to write and publish sustained works on homosexuality, his voice was less a bold venture into *terra incognita* than a foray into territory already conquered by the powerful, homophobic and sexually repressive establishment. He did, however, appropriate discursive terrain for the purposes of resistance, and his daring can be measured by the shocked reaction with which it was met by many in the press (notably the sensationally prudish *Pall Mall Gazette*), as well as by the censorship and expurgation of later editions of his *Nights*,[29] and the destruction of some of his other manuscripts, most notoriously by Isabel.

The *terra incognita* in which Burton mapped the Sotadic Zone was also figurative, concretely geographical – the setting of a specifically utopian/dystopian adventure. When he explains that whereas 'History paints or attempts to paint life as it is', 'Fiction shows or would show us life as it should be, wisely ordered and laid down on fixed lines' (1885, 10.123), Burton hints at the fictional and utopian dimensions of his narrative. He establishes a direct link between himself, as narrator and hero, and the famous utopian/dystopian adventurer Lemuel Gulliver. He mentions, in a telling footnote, that 'Some years ago I was asked by my landlady if ever in the course of my travels I had come across Captain Gulliver' (Ibid., 125). Like Gulliver, Burton travels to a disconnected utopian/dystopian space, off the map; in the preface, he tells how he is transported from the banal surroundings of his English home to the magical setting of the *Nights*, the malleable setting of a utopian adventure. Covering much of the Orient, the geography of the *Nights* was not, of course, a complete blank. There were, however, blank spaces on maps, which Burton was able to fill. First, the gaps left by geographers, anthropologists, translators and others when they neglected to include sexual and other taboo topics in their descriptions. Second, the blank, seemingly empty and timeless geography at the heart of the Sotadic Zone, and at the heart of the *Nights*: the desert. It is in the desert that Burton feels most freedom, not only the freedom to roam around in a world of men and adventure, but also the freedom to imagine, to dream about the past and the future. Edward Said has observed that to Orientalists such as Burton the desert 'appears historically as barren and retarded as it is geographically; the Arabian desert is thus considered to be a locale about which one can make statements regarding the past in exactly the same form (and with the same content) that one makes them regarding the present' (Said 1985, 235). Of course, utopian writers are concerned ultimately with the future. In his *Personal Narrative*, Burton wrote that 'Desert views . . . appeal to the Future, not to the Past: they arouse because they are by no means memorial' (1856, 1.149). Timeless, empty, malleable, literally and metaphorically fluid space, the desert seemed the perfect setting for his utopian adventures. In *terra incognita* such as this, which exists primarily in Burton's geographical imagination and in his writing, the Sotadic Zone is mapped and seems most plausible.

Burton's description of the Sotadic Zone, reminiscent of the four regions in which Gulliver travelled, presents not a static region, but a linear sequence of utopias and dystopias. Narrative is, of course, generally linear, whereas geography is areal, but this does not completely explain the narrative structure in Burton's description of the Sotadic Zone. Burton chooses to represent the Sotadic Zone in linear rather than areal terms. Since he is able to specify geographical coordinates and boundaries with impressive precision, it is noteworthy that Burton neglects to display the Sotadic Zone on a graphic rather than a textual map. His description is like an imaginary journey. The journey begins in Arabia and ancient Greece and then proceeds to other, variously utopian and dystopian regions – moving through Rome and North Africa, and continuing in an easterly direction. 'Proceeding Eastward we reach Egypt, that classical region of all abominations' (1885, 10.224); 'Resuming our way Eastward we find the Sikhs and the Moslems of the Panjab much addicted to Le Vice' (Ibid., 236); 'Passing over to America we find that the Sotadic Zone contains the whole hemisphere' (Ibid., 240), and so on. Thus Burton, who superficially appears to lump all marginal sex together within the Sotadic Zone, actually distinguishes types of sex, and draws lines between forms of good and bad. He distinguishes, most generally, between three types of pederasty – the funny, the grim and the wise (Ibid., 253) – although he goes into detail on subtle geographical variations. Arguing that the love of boys has a 'noble and ideal' side (Ibid., 218), he idealizes Greek and some forms of Arabic pederasty and acknowledges the religious significance of same-sex relationships in Egypt, but criticizes Roman pederasty and scoffs, for example, at the 'systematic bestiality with ducks, goats and other animals' that he claims is common in China (Ibid., 238). By distinguishing between utopias and dystopias, different types of sex, Burton makes the radical claim that same-sex relationships are not necessarily bad – a claim that later gay rights theorists and activists, from radicals such as Edward Carpenter (1984, 257, 260) to antihomophobic doctors such as Kenneth Walker FRCS,[30] were to understand and endorse.

With its graphic sexual imagery, its abstract fleshy surfaces that remind the reader not only of medical textbooks, but also of pornographic literature, Burton's broadly utopian geography is an example of what Steven Marcus calls 'pornotopia' (1966, 269) Printing translated erotica, Burton borrows many of the tricks of pornography: he prints under false or concealed names, behind phoney cover organizations, with false publishers and imprints (Archer 1966). He co-founded a cover organization for the printing of erotica – the Kama-shastra Society – with a membership of just two, himself and F.F. Arbuthnot. The Kama-shastra Society produced books including the *Kama Sutra* and the *Nights* with the false imprint of Benares. The names of translators were concealed, although not particularly well. The *Kama Sutra* was attributed to A.F.F. and B.F.R. – the initials of Arbuthnot and Burton in reverse order. The *Nights* was the first erotic translation Burton published under his own name. It is not clear whether

Burton's pornographic tricks were genuine, or just playful parodies of Britain's burgeoning pornography industry. In either case, the tricks do establish a super-ficial association with pornography, an association that was cemented by Burton's public acquaintance with Monckton Milnes, Fred Hankey, Algernon Charles Swinburne and other well-known collectors of pornography and erotica, an association that is sustained within the books.[31] The style is graphic, the stories replete with sensual, sexual imagery. As Burton explained, 'The gorgeousness is in the imagery not in the language; the words are weak while the sense . . . is strong' (Burton 1885, 10.170). The settings of Burton's *Nights*, like the settings of pornographic literature, are abstractions, historically and geographically vague. The mixture of sexual intimacy and sexual violence – Burton's principal obses-sions include bastinado (beating), castration, genital mutilation and rape[32] – mir-rors that in contemporary English pornography.[33] Yet Burton's interest was not purely voyeuristic, and he was not purely in search of titillation.

Utopian/dystopian writers throughout history have generally used the medium as a form of political criticism, at times when freedom of expression is limited, and Burton was no exception. Burton's pornotopia was a protest against British Victorian homophobia. While many allegations – of prudery, coldness, and general ignorance and horror of their own bodies – hurled at Victorians do not hold up to historical scrutiny, there is little doubt that Victorian Britain was a deeply homophobic society. Michael Mason, who generally defends Victorian sexual attitudes against our caricatures of them, in his detailed history of Victorian sexual attitudes concedes that 'Of the leading prohibitions in the Victorian sexual code only that on homosexuality is not almost universal among other cultures.' (Mason 1994, 4). A rising wave of British homophobia culminated in new legisla-tion, passed in 1885, which led to a series of high-profile trials and prosecutions. In the famous trials of the 1880s and 1890s, Oscar Wilde and others were imprisoned, and most of their homosexual countrymen were forced under-ground. Burton's pornotopia was a specific attack on the amended Criminal Law Amendment Act (1885), which ruled that 'Any male person who in public or private commits or is party to the commission, or attempts to procure the com-mission by any male person, of any act of gross indecency with another male person, shall be guilty of a misdemeanour' (Pearsall 1993, 459). An enforceable two-year imprisonment was the penalty for breaking this law. Burton spelled out his attitude towards this legislation in the 'Terminal Essay', where he argued in the strongest possible terms that sex between men should not be prosecuted. He argued that men who have sex with men 'deserve, not prosecution but the pitiful care of the physician and the study of the psychologist' (1885, 10.209).[34] In the final volume of the *Supplemental Nights*, Burton added a lengthy and combative response to reviewers and the press, particularly the *Pall Mall Gazette* and its editor William T. Stead. Stead was a leading figure in the National Vigilance Association, which promoted the new sexual legislation, and which Burton dis-missed as 'a troop of busybodies' (1886, 6.400). It becomes clear that Burton's homophobia was superficial, his disgusted, pejorative language tongue-in-cheek, lip service to the unwritten rule that homosexuality was to be condemned each time it was mentioned. He subverts his own term for pederasty, disowning it as foreign and meaningless: 'what our neighbours call Le Vice contre nature – as if anything can be contrary to nature which includes all things' (1885, 10.204).[35] The style of Burton's essay on pederasty can be read as a direct response to the political climate of 1885. His emphasis on tangible forms of repression and power and his relative inattention to theories of desire reflect the priorities of the moment, the urgent need to resist new laws, rather than settling down to ponder deeper issues. Thus, when Isabel tried to distance her husband from taints of homosexuality, almost nobody believed her.[36] The public knew little about Richard Burton's sexual biography, but they were never in doubt about his sexual politics, including his opposition to homophobia.

Since the Sotadic Zone is constructed as the setting of a travel narrative, or at least as the distillation of travel narratives, its boundaries are permeable. The traveller who travels from England to Arabia demonstrates that boundaries can be crossed. Travelling, he suggests the possibility of a dialectical relationship between that which lies within the boundaries, and that which lies without. The boundary is less a wall than a connection, a relationship. Burton's conception of boundary, not the reactionary or defensive wall it first appears, is reminiscent of Doreen Massey's reformulation of boundaries, which is also somewhat utopian.

> 'Boundaries' . . . are not necessary for the conceptualisation of a place itself. Definition in this sense does not have to be through simple counterposition to the outside; it can come, in part, precisely through the particularity of linkage *to* that 'outside' which is therefore itself part of what constitutes the place. This helps to get away from the common association between penetrability and vulnerability.
>
> (Massey 1994, 155)

As travellers and adventurers cross boundaries, they sometimes blur distinctions between home and away, centre and periphery, colonizers and colonized, destabilizing rather than reproducing those dualisms (Chambers 1994).

In the 'zone unknown' of Burton's adventure, and in the fantasy of his pornotopia, realism begins to dissolve. Superficially solid ground, which has provided the reader with a point of departure, gives way to something less tangible. Pornotopia, characteristically utopian, is 'no place'. Burton calls his setting 'a mise-en-scène which we suspect can exist and which we know does not' (1885, 10.124). As Mason (1994, 43) shows, in his history of British sexual attitudes, Victorians tended to say one thing about sex while doing another,[37] and Burton was no exception. Cross-checking his claims and anecdotes, and finding inconsistencies, biographers have called him a 'liar' (Brodie 1967, 25, 53). Similarly, geographers and post-colonial critics have accused Burton of misrepresentation.[38] Certainly, he conjured up realistic images, which seemed real, but were not always faithful records of circumstance. And, certainly, his statements about geography and sexuality cannot be taken at face value. But Burton was not a liar. Like other adventure story tellers, Burton made up stories and conjured up settings. He added an 'imaginative varnish' to the *Nights*, which, he explained, 'serves admirably as a foil to the absolute realism of the picture in general' (1885, 10.124). *The Saturday Review* concluded an ambivalent review of the *Nights* by conceding that

> Viewed as a *tout ensemble* in full and complete form, they are a drama of Eastern life, and a Dance of Death made sublime by faith and the highest emotions. . . . They form a phantasmagoria in which archangels and angels, devils and goblins, men of air, of gire, of water, naturally mingle with men of earth; where flying horses and talking fishes are utterly realistic.
>
> (Burton 1886, 6.409)

This reviewer sees that Burton's 'phantasmagoria' are more than lies, even though he is not sure what they are, or where Burton is going. Burton does not say where he is going. Like other adventure and utopian writers, he leads the reader on an entertaining journey towards some important point, but stops short of saying what that point is, and thereby constructs an open-ended space. Burton admits that the *Nights* contain 'fantastic flights of fancy, the wildest improbabilities' and 'the most impossible of impossibilities' (1885, 1.viii). This is true not only of the stories he translates, but of his notes and essays. The Sotadic Zone, which seems real to begin with, does not stand up to cross-examination; its realism begins to dissolve. Having followed Burton to the Sotadic Zone, the reader is then disoriented, forced to find his or her own way. It is at this point that the orthodox map really gives way to the kind of map envisaged by Deleuze and Guattari.

CONCLUSION

Burton charts ambivalent sexual geography – ambivalent geography and ambivalent sexuality – while making a specific political point. He charts sexual geography that seems, on first inspection, to be rigid, static and divided, but turns out to be fluid, open-ended and integrated. Burton's conception of sexuality, which he never quite spells out, is implicit in his ambivalent geography. Biographers who insist on debating whether Burton was homosexual or heterosexual oversimplify and miss the point.[39] Burton's sexual geography was not divided between homosexuals and heterosexuals, nor between homosexuality and heterosexuality. Neither was it a space of bisexuals or bisexuality.[40] Burton eschews such language. Although biographers label him a 'dual man',[41] Burton rejected dualism. His conceptions of sexuality and geography are dynamic and open. Burton does not begin with fixed points and then move between them; he charts dynamic, travelling geography. Burton's decision to use geographical narrative as a medium in which to address sexuality began with his rejection of moral absolutes. In his reading, even the Bible is a historically and geographically specific moral code, not an absolute system (Burton 1885, 10.228). His use of travel, more specifically, expresses disdain for the absoluteness of fixed points and rigid boundaries. His sexual geography is not a world in which 'anything goes', but it is a world in which many rules are lifted and some specific laws repealed.

Burton shows how travel can destabilize dualisms and stereotypes. He has been accused of reproducing Orientalist stereotypes. Kabbani finds it ironic that 'someone with the linguistic and intellectual capacities of Burton should, in the end, have helped only to further confirm the myth of the erotic East' (1986, 66). She argues that Burton, like other Western travellers, 'hardly saw anything at all of the details before them' (Ibid.), and concludes that '[t]he East became codified and static in ways that were final; no deeper perception was permissible, nor indeed possible given the weighty heritage of prejudice' (Ibid., 139). According to this view, the crudely Orientalist Burton reproduced stereotyped constructions of here and there, us and them. Said generalized that 'the details of Oriental life serve merely to reassert the Orientalness of the subject and the Westernness of the observer' (1985, 247). I have shown, however, that Burton was never content to reproduce accepted ideas, nor to reproduce hegemonic constructions of geography and identity. On the subject of empire, he was not content merely to reproduce; he wanted to build a bigger and stronger empire, with the use of force where necessary. On the subject of sex, he was equally willing to dissent from the contemporary mainstream, if this time to follow a radically permissive line. In each case, Burton charted space in which to envisage change. He mapped a world where nothing stands still, a world where boundaries are constantly transgressed, and where there is no place for the conservative.

ACKNOWLEDGEMENTS

I am grateful to Derek Gregory for introducing me to Richard Burton, and to Diane Watt and Vaughan Cummins for commenting on earlier drafts of this paper.

NOTES

1 See also Bell and Valentine (1995) and Bleys (1996).
2 I develop this idea of mapping in greater detail in *Mapping Men and Empire: A geography of adventure* (1997).
3 The title page continues, 'With Introduction Explanatory Notes on the Manners and Customs of Moslem Men and a Terminal Essay upon the History of the Nights. Printed by the Kama-shastra Society For Private Subscribers Only. 10 vols.'
4 Burton was the author of over forty published books although these, among his earliest works, were his most successful travel books. They stayed in print almost continuously. Burton's *Pilgrimage to El-Medinah and Meccah*, for example, was reprinted, and repackaged in at least five different editions between 1855 and 1893, when Isabel Burton issued a 'Memorial Edition' (Penzer 1923).
5 Spurred on by the combative and adversarial nature of their dispute, which centred around their geographical explorations, Burton and Speke both made some implausible and overstated claims. As Brodie (1967, 224) puts it, Burton was

'guilty of . . . irresponsible geography-writing'. Speke died shortly before he was to have met Burton in a public debate. Brodie interprets his suicide as a final, 'supreme act of hate' towards his former travelling companion and, perhaps, lover (Ibid., 226).

6 Burton's appendix to *First Footsteps in East Africa*, entitled 'Brief Description of Certain Peculiar Customs', which described acts of adultery and positions of Somali love, proved too much for the publisher, who ordered the appendix to be ripped out (Brodie 1967, 110).

7 Foucault (1978) speaks of the Victorian search for 'truth' about sex.

8 This has been explored by others, notably Aldrich (1993).

9 Originally ten volumes, followed by six 'supplemental nights', which were later divided into seven volumes. Penzer (1923, 135) lists subsequent editions including Lady Burton's (6 volumes, 1886–8), Smithers' (12 and 13 volumes, 1894 and 1897), Denver Burton Society (16 volumes, 1900–1), Burton Club (17 volumes, 1903–4) and Burton Club 'Catch word' (17 volumes, 1905–20).

10 The distinction between publishing and printing, previously blurred, was relatively sharp by 1885.

11 Burton contributed to Payne's translation, which was published in an edition of 500 copies by subscription, and he dedicated a volume of his own translation to Payne.

12 Walker (1964) prefers not to use the term 'pederasty', given its purely sexual, and negative, connotations today. He replaces the language of pederasty with that of homosexuality.

13 Extracts from Havelock Ellis (*Sexual Inversion: Studies in the Psychology of Sex*) and John Addington Symonds (reviews) are reproduced in Penzer (n.d.). Symonds, author of an early and favourable review of Burton's *Nights*, was more impressed by Burton's translations and notes than by his most general remarks and theories. Burton reproduced and responded to Symonds' review of the *Nights* in *Supplemental Nights* (1886, 6.406, 412).

14 Although, as Burton explains in a footnote, 'As this feminine perversion is only glanced at in the *Nights* I need hardly enlarge upon the subject' (Burton 1885, 10.209).

15 The erotic fantasy is identified by Kabbani (1986), with a homoerotic dimension explored by Boone (1995).

16 Burton co-founded the Cannibal Club (1863) and the Anthropological Society of London (1873), vehicles for 'learned debauchery' and sexual ethnology respectively (Brodie 1967; see also Archer 1966).

17 Sex did not emerge as a preoccupation of mainstream anthropologists until after Bronislaw Malinkowski's *The Sexual Life of Savages* was published, in 1929, with a foreword by Havelock Ellis (Kulick 1995).

18 Unpaginated one-page 'Memorandum' sent by Burton with Volume 1 of the *Nights* in 1885.

19 Assad (1964) suggests that Burton adheres closely to the text of the original Nights, but adds much that is his own in the notes. To Kabbani (1986) even the 'translation' is Burton's creation.

20 In a letter to *The Academy*, dated 6 March 1886, Isabel Burton wrote, 'I have not read, nor do I intend to read, my husband's *Arabian Nights*.'

21 While he described Payne's *Nights*, of which 500 copies were printed, as 'caviare to the general – practically unprocurable' (1885, 1.xiii), Burton issued just 1,000 copies of his own version.

22 Burton was probably familiar with the relatively new concept of homosexuality, which dates to 1869. He is said to have discussed the work of Karl Heinrich Ulrichs, who coined the term homosexuality, in the manuscript of his revised translation of *The Scented Garden*, which contained a lengthy discourse on same-sex love.

23 McLynn writes that 'There was a certain ambiguity in his sexuality and an ambivalence in his study of homosexuality, prostitution, pederasty, castration and infibulation. He dallied with Indian and Persian mistresses, experimented with black and African women in Africa and Asia, and whoremongered his way through the brothels of Paris' (1990: 8).

24 Burton's *Nights* were originally told to women as well as men in the Arab world.

25 In a typical footnote, he writes, 'For full details I must refer readers to my 'Personal Narrative of a Pilgrimage to El-Medinah and Meccah' . . . I shall have often to refer to it' (Burton 1885, 1.28).

26 Burton pointed out that literary works by Shakespeare, Swift, Rabelais and others were replete with sexual references, which would probably have been considered scandalous if printed in other media. Barret-Ducrocq (1991, 1) notes that Victorian Britain 'had nothing to say on sexual matters but left them to the professionals: medical specialist, pornographer and prostitute'.

27 Principal agents of the 1880s purity campaigns include William Coote's National Vigilance Association, Ellice Hopkins' Christian Feminist movement and W.T. Stead's *Pall Mall Gazette* (Bristow 1977).

28 Mason devotes very little attention to homosexuality in *The Making of Victorian Sexuality, Volume One*, because, he argues, 'the evidence about homosexuality in the period' before 1880 is 'extremely meagre' (1994, 6). From the mid-1880s to the end of the century, increased attention was drawn to homosexuality, according to Pearsall (1993, 459).

29 For example, the twelve-volume 1894 edition, edited by Leonard C. Smithers. Smithers notes that, while 'greater latitude is properly allowable' owing to the book's 'scientific and ethnographical' importance, the 'extreme grossness' of some parts renders them unprintable (1.viii). Smithers omits most of the terminal essay on pederasty, noting only that 'It has been deemed necessary to omit from this volume the Article on Pederasty' (8.185).

30 At a time when male homosexuality was still illegal in Britain, Walker (1964) saw fit to edit an anthology of Burton's 'extremely interesting' (18) essays on sexuality, and to endorse Burton's view that 'homosexuals should be the responsibility of the doctor and not of the judge' (20).

31 Burton dedicated a volume of the *Nights* to Monckton Milnes, a well-known collector of erotica. He cites Pisanus Fraxi's virtually definitive bibliographies of pornography and sexual literature: *Index Librorum Prohibitorum* (London, 1877), *Centuria Librorum Absconditorum* (1879) and *Catena Librorum Tacendorum* (1885) (Burton 1885, 10.252). Fraxi was also known as Henry Spencer Ashby (Marcus 1966). Burton is reputed to have beaten the camp poet, Swinburne, on a number of occasions, with the latter's enthusiastic consent (Brodie 1967, 197).

32 Burton maintained his obsession with these themes throughout his writing career. Some of his works, such as *A Mission to Gelele King of Dahome* (1864), are almost entirely devoted to them.

33 Marcus (1966) identifies flagellation as an obsession of British Victorian pornography.

34 This sounds equivocal today, but Burton's argument was radical in 1885.

35 He adds, in a footnote (same page), 'Amongst the wiser ancients sinning contra naturam was not marrying and begetting children.' Burton himself was guilty of the second of these 'sins', since he had no children.

36 Isabel's attempts to distance her husband from homosexuality, which extended to burning his papers and MS. of *The Scented Garden*, to explaining her actions in letters to the press, backfired since her denials began to appear incriminating (Brodie 1967, 329). She then wrote a lengthy biography of her husband, intended

to retrieve his memory. In the biography, Richard is presented as heterosexual, loyal in marriage and sexually pure (Burton, I. 1893).

37 Mason notes the common anthropological finding that 'a considerable discrepancy between sexual codes and sexual activities is common in human societies', and finds Victorian Britain no exception to this.

38 Kabbani (1986, 10) generalizes that Western travel writers 'misrepresented' the East. Said (1985, 273) avoids the problematic language of representation and misrepresentation and argues that Orientalist discourse 'is not a misrepresentation of some Oriental essence'.

39 For example, McLynn (1990) argues the case that Burton was not homosexual, while Archer (1966, 19) attempts to show that he was, citing Burton's interest in swords as evidence of repressed homosexuality! Aldrich (1993, 173), seeking impossible middle ground, says only that Burton 'may have been homosexual'.

40 The language of bisexuality is out of place in the Arab world (the idea of heterosexual, bisexual and homosexual identities is geographically and historically limited, arguably to the West and to the modern period), although Westerners (such as Gollain 1996) often persist in using it.

41 For example, McLynn (1990, 2). Brodie (1967, 175) comments on Burton's 'bi-polarity'.

REFERENCES

Aldrich, Robert (1993) *The Seduction of the Mediterranean*, London: Routledge.

Alpers, S. (1983) *The Art of Describing*, Chicago: University of Chicago Press.

Archer, W.G. (ed.) (1966) *The Kama Sutra of Vatsyayana*, trans. R.F. Burton, London: Allen and Unwin.

Assad, Thomas J. (1964) *3 Victorian Travellers*, London: Routledge.

Barret-Ducrocq, Françoise (1991) *Love in the Time of Victoria*, trans. J. Howe, London: Verso.

Bell, David (1995) '[screw]ING GEOGRAPHY', *Environment and Planning D, Society and Space* 13, pp. 127–31.

Bell, David and Valentine, Gill (eds) (1995) *Mapping Desire: Geographies of sexualities*, London: Routledge.

Bleys, Rudi C. (1996) *The Geography of Perversion: Male-to-male sexual behaviour outside the West and the ethnographic imagination 1750–1918*, London: Cassell.

Boone, Joseph (1995) 'Vacation cruises; or, the homoerotics of Orientalism', *Publications of the Modern Language Association* 110, pp. 89–107.

Bristow, Edward (1977) *Vice and Vigilance: Purity movements in Britain since 1700*, Dublin: Gill and Macmillan.

Brodie, Fawn M. (1967) *The Devil Drives: A life of Sir Richard Burton*, London: Eyre & Spottiswoode.

Burton, Isabel (1893) *The Life of Captain Sir Richard F. Burton*, London: Chapman & Hall.

Burton, R.F. (1855–6) *Personal Narrative of a Pilgrimage to El-Medinah and Meccah*, London: Longman.

—— (1856) *First Footsteps in East Africa; or, An Exploration of Harar*, London: Longman.

Burton, R.F. (1860) *The Lake Regions of Central Africa, A Picture of Exploration*, London: Longman.

—— (1864) *A Mission to Gelele, King of Dahome*, London: Tinsley.

—— (1883) *Kama Sutra of Vatsyayana*, Benares: Kama-shastra Society.

—— (1885) *Ananga-Ranga or the Hindu Art of Love*, Benares: Kama-shastra Society.

—— (1885–6) *Plain and Literal Translation of the Arabian Nights' Entertainments or The Book of a Thousand Nights and a Night*, Benares: Kama-shastra Society.

—— (1886) *Supplemented Nights. To the Book of the Thousand Nights and a Night. With Notes Anthropological and Explanatory*, Benares: Kama-shastra Society.

Campbell, J. (1949) *The Hero with a Thousand Faces*, New York: Bollingen/Pantheon.

Caracciolo, Peter L. (ed.) (1988) *The Arabian Nights in English Literature*, Basingstoke: Macmillan.

Carpenter, Edward (1984) *Edward Carpenter. Selected Writings Volume One: Sex*, London: GMP.

Carter, P. (1988) *The Road to Botany Bay: An exploration of landscape and history*, New York: Alfred Knopf.

Chambers, I. (ed.) (1994) *Migrancy, Culture, Identity*, London: Routledge.

Deleuze, G. and Guattari, F. (1988) *A Thousand Plateaus: Capitalism and schizophrenia*, London: Athlone.

Duncan, J. and Duncan, N. (1988) '(Re)reading the landscape', *Environment and Planning D, Society and Space* 6, pp. 117–26.

Foucault, Michel (1978) *The History of Sexuality, Volume 1*, Harmondsworth: Penguin.

Frye, Northrop (1990) *The Anatomy of Criticism*, Harmondsworth: Penguin.

Gay, Peter (1986) *The Education of the Senses*, Oxford and New York: Oxford University Press.

Gollain, F. (1996) 'Bisexuality in the Arab world', in Sharon Rose (ed.), *Bisexual Horizons: Politics, histories, lives*, New York: Lawrence & Wishart, pp. 58–61.

Harley, J.B. (1992) 'Deconstructing the map', in T. Barnes and J. Duncan (eds), *Writing Worlds*, London: Routledge, pp. 231–47.

Hyam, Ronald (1990) *Empire and Sexuality: The British experience*, Manchester: Manchester University Press.

Kabbani, Rana (1986) *Imperial Fictions: Europe's myths of Orient*, London: Pandora.

Kulick, Don (1995) 'Wild oats in the open', *Times Higher Education Supplement* 1200 (3 Nov.), p. 19.

McLynn, Frank (1990) *Of No Country: An anthology of the works of Richard Burton*, London: Scribners.

Malinkowski, B. (1929) *The Sexual Life of Savages*, London: Routledge.

Marcus, Steven (1966) *The Other Victorians: A study of sexuality and pornography in mid-nineteenth century England*, London: Weidenfeld and Nicolson.

Mason, M. (1994) *The Making of Victorian Sexuality*, Oxford: Oxford University Press.

Massey, Doreen (1994) 'A global sense of place', in D. Massey (ed.), *Space, Place and Gender*, Cambridge: Polity.

Morris, Jan (1996) 'In a family embrace', *The Guardian*, Saturday 9 March, p. 29.

Moussa-Mahmoud, Fatma (1988) 'English travellers and the Arabian Nights', in Peter L. Caracciolo (ed.), *The Arabian Nights in English Literature*, Basingstoke: Macmillan, pp. 95–110.

Pearsall, R. (1993) *The Worm in the Bud: The world of Victorian sexuality*, London: Pimlico.

Penzer, Norman (1923) *Annotated Bibliography of Sir Richard Burton*, London: Philpot.

—— (ed.) (n.d.) *Anthropological Notes on the Sotadic Zone*, New York: Falstaff.

Phillips, Richard (1997) *Mapping Men and Empire: A geography of adventure*, London: Routledge.

Said, Edward (1985) *Orientalism*, Harmondsworth: Penguin.

Smith, N. (1994) 'Geography, empire and social theory', *Progress in Human Geography* 18, 491–500.

Walker, Kenneth (ed.) (1964) *Love, War and Fancy: The customs and manners of the East from writings of the Arabian Nights*, London: William Kimber.

Zweig, Paul (1974) *The Adventurer*, London: Dent.

5   *The Flight from Lucknow*

British women travelling and writing home, 1857–8

Alison Blunt

A number of paintings exhibited at the Royal Academy in London in the summer of 1858 took as their subject the recently suppressed Indian 'mutiny'/uprising. This unrest had broken out in Central and Northern India and lasted from May 1857 to June 1858.[1] The inclusion of such paintings at the Royal Academy – together with daily newspaper accounts and Parliamentary debates – reflected the unprecedented level of public attention in Britain being paid to events in India. One focus of attention was the position of British women in India, with sensationalist accounts of their deaths and barely veiled (but unsupported) hints at their violation resulting in particularly bloodthirsty cries for vengeance.[2] Popular interest about the place of British women in India was also reflected in the Royal Academy exhibition, with paintings by Edward Armitage, Edgar George Papworth, Joseph Noel Paton and Abraham Solomon all depicting British women in the 'mutiny' (Harrington 1993). One of these paintings, *The Flight from Lucknow* by Abraham Solomon (Figure 1), provides the title and subject of this chapter.

The Flight from Lucknow represents the evacuation of Lucknow in November 1857. Together with the rest of the British population in Lucknow, 240 women were confined to the Residency compound from June to November 1857.[3] The majority of these women were married to soldiers, but 69 'ladies' were related to officers or officials (Innes 1895).[4] A number of these 'ladies' recorded their experiences of living under siege in diaries and letters, some of which were subsequently published.[5] In September, an unsuccessful 'relief' provided reinforcements. Forces sent from Britain, under the command of Sir Colin Campbell, relieved Lucknow for the second time on 17 November. This was followed by the evacuation of Lucknow, first by the injured, and then, on 19 November, by British women and children. This evacuation was followed by the withdrawal of all British troops from Lucknow by 23 November, although fighting continued until Lucknow was recaptured by the British in March 1858 (Hibbert 1978). As well as recording their lives under siege, the diaries and letters written by women also described their evacuation from the Lucknow Residency and their three-month journey by foot, rail and steamer to Calcutta.

The writings of British women travelling from Lucknow to Calcutta from November 1857 to January 1858 represent collective rather than individual

Figure 1 The Flight from Lucknow, by Abraham Solomon, 1858. (Courtesy of the Museum and Art Gallery of Leicester.)

travel, which took place under forced rather than voluntary conditions. As such, my interest in their diaries and letters relates to but also differs from other studies of imperial travel writing by women (e.g. Blunt 1994; Blunt and Rose 1994; Mills 1991). The spatial extent of the British empire in the nineteenth and early twentieth centuries enabled middle-class British women to travel more widely than fore, and the travels and writings of many of these women are attracting an ising amount of critical attention. It has been shown not only that British nen as well as men were imperial subjects but also that their imperial and ndered subjectivity was closely influenced by their experiences of travel. Work on British women travellers has focused on their ability to transgress the confines of 'home' in social as well as spatial terms. The travels and writings of individual women suggest that they were empowered to travel and transgress in the context of imperialism while away from the feminized domesticity of living 'at home' (Blunt 1994; Mills 1996). However, the majority of British women who travelled in the context of imperialism did so to set up home both with and for their families and in this chapter I want to focus on the writings of a group of British women who were living in India rather than on individuals who have come to be represented as travel writers. I want to argue that British women played often ambivalent roles as both domestic and imperial subjects travelling and living in imperial places. While I agree with Anne McClintock that 'the cultural history of imperialism cannot be understood without a theory of domestic space and gender power' (McClintock 1995, 133),[6] I would add that the interconnections between imperialism, domestic space and gender power cannot be understood without a focus on travel and representations of travel that underpinned imperial rule both over space and in particular places. Imperial power and legitimation relied not only on imaginative geographies of 'other' places (Said 1979) but also on imaginative geographies of 'home', both between Britain and India and within India itself. Just as the spaces of home and away were more blurred than distinct through imperial travel, so too were the domestic and imperial subjectivities of British women both travelling and living in imperial places.

In this chapter, I want to focus on British women travelling and writing home at a time of imperial conflict to examine the ways in which they not only represented but also began to reconstitute their domestic and imperial subjectivity as they moved away from Lucknow. I will consider how imaginative geographies of home and away on different scales helped to influence the domestic and imperial subjectivity of British women as they travelled from the dangers of Lucknow and Cawnpore towards the safety of Allahabad and Calcutta. Before doing so, however, I want to begin by discussing the ambivalent imperial representations not only of the evacuation of Lucknow but also of the place of British women in the evacuation.

THE FLIGHT FROM LUCKNOW

Solomon's painting depicts a group of women and children fleeing a group of burning buildings. Two white women at the front seem particularly out of place. They are dressed in expensive, highly decorative clothes, with one supporting and leading the other while looking anxiously back. They are followed by an Indian ayah, holding a sleeping white child, who is followed by another child, leading more women on. While the painting conveys a clear sense of escape, danger and anxiety, it also suggests a degree of comfort, care and protection both between the white women and between the ayah and white child.

In marked contrast to popular knowledge about the siege of Lucknow, reactions to Solomon's painting were strangely placeless. The siege of Lucknow continued to shape the British imperial imagination over many years following the 'mutiny'. For example, *The Times* stated in 1930 that 'probably no achievement in British history stirs the blood of Englishmen more deeply than the defence of Lucknow' (18 February 1930). Indeed, after the recapture of the Residency at Lucknow in March 1858, the Union Jack was lowered from its tower for the first and last time on 15 August 1947, the date of Indian Independence. This was the only Union Jack in the British Empire to fly day and night and when it was finally lowered, the *Illustrated London News* reported the event as 'probably the most poignant flag ceremony of the day'.[7] However, the evacuation of Lucknow occupied an ambivalent place in such imperial memories. For example, it was the date of the first, unsuccessful 'relief' – 25 September – that came to be known as Lucknow Day in the years following the 'mutiny', and which was marked by an annual dinner of survivors.[8] Even representations of the evacuation of Lucknow came to be appreciated in generic rather than place-specific terms. For example, Solomon's painting came to be known as *The Flight*, and its specific reference to Lucknow was often obscured. A critic in the *Athenaeum* wrote that 'Some English ladies are escaping from some Indian massacre', while a critic in the *Art Journal* agreed that 'The scene is India, and the fugitives are a party of our countrywomen flying in terror from a burning city'. As the latter critic continued, 'Any episode of this kind cannot be far from the truth, since these flights have occurred too frequently.'[9]

The place of British women in the evacuation of Lucknow was represented in ambivalent ways in both contemporary reports and histories of the 'mutiny'. Their rescue was represented in heroic terms, as they had been saved from the fate of women and children in other places, particularly at Cawnpore.[10] At the time of the evacuation, Sir Colin Campbell wrote that 'The persevering constancy of this small garrison, under the watchful command of the Brigadier, has, under Providence, been the means of adding to the prestige of the British army and of preserving the honour and lives of our countrywomen.'[11] It was also only once British women had been evacuated that their lives under siege could be represented. For example, as the Calcutta correspondent of the *Daily News* wrote, 'It was by no means an easy matter to move the ladies out of the place in which they had so long borne up against privations and danger with more than heroic fortitude',[12] and the *Illustrated London News* reported that 'The privations endured by the heroic garrison, and particularly by the ladies, were fearful' (9 January 1858). Newspaper reports about such women increased as they travelled closer to Allahabad and Calcutta where, on their arrival, they were received and represented as heroines. It was only possible to represent the evacuated women as heroines away from Lucknow, where they had been out of place, under threat and suffering privations and dangers. Both Allahabad and Calcutta were a safe distance away from Lucknow and offered imperial and domestic security.

One effect of representing the evacuated women as heroines was to divert attention away from the fighting at Lucknow that continued for the next four months. In a short period of time, however, the status of the Lucknow heroines became open to question. For example, William Howard Russell, the *Times* correspondent, wrote in his diary (30 January 1858) of meeting a woman evacuated from Lucknow and living in a temporary home in Calcutta. He was shocked to hear that 'there was a good deal of etiquette about visiting and speaking in the garrison! Strange, whilst cannon-shot and shell were rending the walls about their ears – whilst disease was knocking at the door of every room, that those artificial rules of life still exercised their force' (Russell 1957, 14). In 1859, an article in the *Calcutta Review* was critical of Lucknow heroines, stating that 'it is humbling to reflect that some of the Lucknow ladies have since been polking to the tune of the "Relief of Lucknow". The fact is, great trials do not alter the character; they only manifest and to a certain degree modify it' (September 1859). Their role as heroines was only possible once they had been evacuated from Lucknow, but this role remained unquestioned for only a short period of time. Once the political expedience of representing the evacuated women as heroines had disappeared, so too did their status.

Another commentator who represented the British women evacuated from Lucknow in more critical ways was Friedrich Engels. In an article for the New York *Daily Tribune* in April 1858, he wrote about the withdrawal from Lucknow to an entrenched camp at Cawnpore, from which Campbell based his campaign to recapture Lucknow. Before continuing with this campaign, however,

> he had another task to perform before he thought it safe to move – a task the attempting of which at once distinguishes him from almost all preceding Indian commanders. He would have no women loitering about the camp. He had had quite enough of the 'heroines' at Lucknow, and on the march to Cawnpore; they had considered it quite natural that the movements of the army, as had always been the case in India, should be subordinate to their fancies and their comfort. No sooner had Campbell reached Cawnpore than he sent the whole interesting and troublesome community to Allahabad, out of his way.
>
> (Engels 1858)

In this article, Engels not only questioned the status of the 'Lucknow heroines' but also the status of an army that would, without Campbell's leadership, be subservient to their wishes. Indeed, as Maria Germon wrote in her diary, 'no doubt it was an anxious time for [Sir Colin] though we did grumble as he did not appear to think much of our comfort' (Germon 1957, 131).[13] Expressing much the same sentiments as Engels, but in a different tone, Julia Inglis (Figure 2), the wife of the man in command of the Lucknow defence, recorded a conversation with Campbell on the day of the evacuation. As she wrote in her diary, 'he was very kind in his manner, and talked about us as dear creatures, meaning the ladies; at the same time, I knew he was wishing us very far away, and no wonder!'

Figure 2 'Colonel Inglis, the Commandant at Lucknow; and Mrs Inglis and Family', *Illustrated London News*, 28 November 1857.

(1892, 200). I want now to focus on the writings of Julia Inglis, Maria Germon, and other British women who made up the 'interesting and troublesome community' travelling and writing home from Lucknow.

TRAVELLING AND WRITING HOME

The diaries by British women at Lucknow were written to record their daily lives both for themselves and for their families and friends at 'home' in Britain. As Katherine Harris wrote,

> I have kept a rough sort of journal during the whole siege, often written under the greatest difficulties – part of the time with a child in my arms or asleep in my lap; but I persevered, because I knew if we survived you would like to live our siege life over in imagination, and the little details would interest you; besides the comfort of talking to you.
> (Letter to her family from Allahabad, 14 December 1857 [Harris 1858, iii])[14]

On their publication in 1858, the diaries of Katherine Harris, Katherine Bartrum and Adelaide Case reached an audience beyond the family and friends for whom they had been initially intended.[15] Each author was keen to explain her reasons for publication and stressed that publication had been suggested by friends.[16] Adelaide Case wrote that she hoped to supplement official despatches about Lucknow with her account of daily life under siege. As she said, 'I have not attempted, by subsequent additions, to produce effect, or to aim at glowing descriptions, but have given it as it was written, in the simple narrative form, which the dangers and privations of the siege alone permitted' (Case 1858, iii–iv), suggesting that her original diary entries could most effectively convey the immediacy and authenticity of her experiences. In a similar way, Katherine Bartrum wrote that her diary represented her personal experiences of daily life and domesticity under siege, stating in self-deprecating tones that

> It is not the wish of the writer of this little Volume, any more than it is in her power, to draw, in glowing colours, a picture of sights and scenes through which it has been her lot to pass, but merely, at the desire of her friends, to give in simple truthfulness a detail of those domestic occurrences which fell immediately under her own observation during the siege of Lucknow.[17]
>
> (Bartrum 1858, preface)

Throughout their diaries, British women interpreted the significance of daily life in domestic more than in imperial terms.[18] At the beginning of the siege, while their Indian servants remained, there was some continuity not only of home life but also of imperial rule from the perspective of British women diarists. Within a few days, however, all of the diarists recorded the desertion of Indian servants in tones of great indignation. Katherine Harris wrote, for example, that 'Their impudence is beyond bounds: they are losing even the semblance of respect' (1858, 47). For the first time, these women had to make tea, clean, wash their clothes and, occasionally, cook, although the wives of British soldiers were usually employed for this purpose. As Katherine Bartrum wrote in June, 'All our servants have deserted us, and now our trials have begun in earnest . . . how we are to manage now, I cannot tell' (1858, 21).

For these women, imperial power was challenged most directly in a domestic sphere because this is where their own imperial subjectivity was challenged. By employing Indian servants, constructions of the racial superiority of British women enabled them to share in imperial power on a domestic scale, establishing what Rosemary Marangoly George (1994) has called their empire in the home. However, when they had to do domestic work themselves, their constructions of racial superiority were destabilized and the basis of imperial legitimation seems threatened in the very areas – home and daily life – on which it had previously relied. In his history of the 'mutiny', Sir John Kaye wrote that 'our women were not dishonoured, save that they were made to feel their servitude' (1876, 354) and, as Jenny Sharpe argues, 'the rebels had unsettled a colonial order to the degree of reversing its hierarchy of mastery and servitude' (Sharpe 1993, 65).[19] However, such a reversal was clearly gender and class specific as it was only middle-class women – the sixty-nine 'ladies' – who were made to feel an unaccustomed servitude. If British men had been positioned in this way, imperial self-legitimation would have been completely compromised and, for the wives of British soldiers, who made up the majority of British women at Lucknow, such servitude was nothing new.[20]

As their diaries reveal, British women at Lucknow experienced a crisis of imperial rule on a domestic scale. In this chapter, I want to focus on the ways in which British women not only represented but also reconstituted their domestic and imperial subjectivity as they travelled away from Lucknow. Rather than view the 'domestic occurrences' experienced by women like Katherine Bartrum as separate from 'the sights and scenes' through which they passed, I want to reveal some of their connections. To do so, I will focus on the ways in which British women represented, first, the dangers of travelling away from Lucknow and visiting Cawnpore and, second, the ways in which they began to reconstruct home while living in the security of Allahabad and Calcutta.

TRAVELLING AWAY

On 17 November the British women diarists recorded their shock at learning that they were to be evacuated from Lucknow the following day. Katherine Harris wrote, for example, that 'We were astounded this morning after prayers by the news that *tomorrow night* this place is to be evacuated. We are all to leave it, with only as much of our worldly goods as we can carry in our hands. I feel utterly bewildered' (1858, 160).[21] In the event, their departure was delayed by another day because of the need to evacuate the sick and wounded first. In their diverse representations of the evacuation, British women recorded joy, regret and anxiety at leaving Lucknow. Katherine Harris expressed her surprise and concern at the evacuation, writing that 'It seems such an extraordinary step, after holding the garrison for so long; no one ever dreamed for a moment of such a measure as evacuating Oude now. I trust it is all for the best. If we live to reach Calcutta, we shall be in a state of destitution' (Ibid., 160–1).

Julia Inglis also wrote about the strategic implications of the evacuation, perceiving her place at Lucknow as part of the British defence. As she wrote, 'We were . . . truly grieved to think of abandoning the place we had held so long with a small force, now that it seemed to us we could have driven the enemy completely out of Lucknow, reestablished our supremacy, and marched out triumphantly' (1892, 197). In contrast, rather than perceive herself as part of an heroic Lucknow defence, Ann Ellen Huxham wrote of her 'deep feelings of joy and gratitude' at being saved by other heroic defenders. As she left the Residency compound, 'many of our dear good soldiers who were standing there on duty and who had risked their lives for us, accosted us with kind words such as "God bless you, we are so glad to have saved you", and as we passed onwards we thanked them heartily' (Huxham n.d.).

In their accounts of leaving Lucknow, however, most British women wrote about the material, more domestic concerns of how to pack and transport their 'worldly goods' rather than the strategic implications of the evacuation. While some women were told that one camel per person would bear their possessions, others were told that they could only take what they could wear and carry. Colina Brydon, for example, was able to take all her possessions except her harp, although three days after leaving Lucknow she had procured a carriage and sent four servants back, 'through a good deal of firing', to fetch it (Brydon 1978, 69). Other women made bonfires of their property to prevent it falling into rebel hands (Inglis 1892), and sewed their valuables into their clothes. Maria Germon, for example, sewed her mother's fish knife and fork into her skirt, filled her pockets with her jewellery and journal, and wore a bag with her lace sewn up inside. She also wore as many clothes as possible, including 'four flannel waist-coats, three pairs of stockings, three chemises, three drawers, one flannel and four white petticoats, my pink flannel dressing-gown skirt, plaid jacket and over all my cloth dress and jacket' (Germon 1957, 120–1).[22]

In their diaries, most British women wrote about the personal logistics of packing and leaving Lucknow rather than the strategic implications of the evacuation. Furthermore, as they left Lucknow, few British women commented on the areas of fighting through which they travelled. Adelaide Case was the only woman to record her impressions of Lucknow as she left the city:

> The scene of ruin, devastation, and misery which presented itself to our eyes when we got out I *never, never* shall forget. To describe it would be impossible; but the horrors of war presented themselves with full force in the mass of shattered buildings and dilapidated gateways through which we passed.
>
> (Case 1858, 288–9)

In their accounts of leaving Lucknow, however, most British women appear largely detached from 'the horrors of war'. This detachment became most clearly apparent as, after walking for an hour, they assembled in the gardens at Sikandar Bagh, which had been the site of heavy fighting three days before.[23] The advancing British forces had killed over a thousand Indians, whose bodies had been barely covered with soil by the time the evacuated women arrived. Despite this, Julia Inglis was the only diarist to comment on the proximity of one of 'the horrors of war', writing that 'Nearly 1200 of them had been cut to pieces, no quarter being asked or granted. Their bodies had just been covered over with earth, and it sickened me to feel they were so near us' (1892, 200). Other women diarists described Sikandar Bagh as the place where they ate 'a regular feast' of bread and butter, beef and chicken. These 'long untasted luxuries' provided the first experience of freedom since living under siege for women who continued to represent an imperial crisis in largely domestic terms (Harris 1858, 163).[24]

Their representations of the crisis began to change, however, when the British women reached the Dilkusha Palace at midnight on 19 November and received letters from Britain that had been accumulating for the past five months.[25] In their writings about leaving Lucknow, letters from 'home' prompted British women diarists to an unprecedented level of self-reflection about the crisis, not only on a personal level but also on an imperial level. On a personal level, Katherine

Bartrum received letters from her husband before his death in September, and from his mother, who did not know of his death, prompting her to write: 'How changed the scene from this day three years ago, when all looked so bright and fair; now I am alone, and there are few in this strange land to care for me' (1858, 57). Letters from 'home' made British women aware for the first time of public as well as familial concern about their place at the centre of an imperial crisis. Adelaide Case wrote that 'The anxiety of England is heartrending to think of, and public sympathy seems indeed to be bestowed upon us to the very highest degree' (1858, 294), while Julia Inglis reflected on the imperial as well as personal implications of the crisis: 'how many, many sad hearts and homes there must be in England just now; and really at present one cannot see an end to our troubles. The whole of Bengal is such an unsettled state that no one can tell when or where a fresh disturbance may break out' (1858, 29–30). During the siege, British women represented the crisis in domestic rather than imperial terms and continued to do so for the first few days after leaving Lucknow. However, once they received letters from 'home' in Britain, the domestic and imperial subjectivity of these women became more closely intertwined. The public and familial concern expressed in such letters prompted British women to examine their position at the centre of an imperial as well as a personal and domestic crisis.

The hasty evacuation of British women and children from Lucknow, followed by the withdrawal of all British forces, suggest that even after the second 'relief' British women continued to be at the centre of an imperial crisis. However, their representations not only of such an imperial crisis but also of their place within it were often ambivalent. Most British women wrote about leaving Lucknow in personal and domestic terms that seem largely detached from the imperial, strategic context in which they were living and travelling. For example, while they were all concerned with packing and transporting their material possessions and with the dietary luxuries of freedom, few British women wrote about the destruction of Lucknow and the recent battle at Sikandar Bagh. It was only when they received letters from Britain that they began to represent the imperial as well as the personal and domestic implications of the siege of Lucknow. In the first few days spent travelling away from their confinement at Lucknow, British women diarists wrote little about the places through which they passed and their subjectivity seems equally confined to personal and domestic rather than imperial concerns. However, on receiving letters from 'home', they began to represent their position in different ways, describing the crisis in imperial as well as domestic terms. Clearly, the imperial and domestic subjectivities of British women diarists were reconstituted not only by their material travel away from the dangers of Lucknow towards the safety of Allahabad and Calcutta, but also by their vicarious travel 'home' to Britain. While such self-reflective representations were triggered by receiving letters from Britain, they were further extended by visiting Cawnpore.

The journey to Cawnpore was slow, hot and dusty. Maria Germon wrote, 'Never shall I forget the scene – as far as the eye could search on all sides were strings of vehicles, elephants, camels, etc. The dust was overpowering'

(1957, 127), while Katherine Bartrum recorded that the confusion of the march was 'perfectly indescribable' (1858, 58), Cawnpore continued to be a site of conflict and, as they crossed the Ganges at night and in silence on 30 November, the sound of gunfire was clearly audible. Despite being closely besieged for five months, Katherine Harris wrote that 'I never, during the whole siege, more thoroughly realised such an extreme sense of nearly-impending danger, and how very close death might be: one felt as if the very next instant perhaps might be one's last. I shall never forget crossing that river' (1858, 178).

In contrast to their previous detachment from both the conflict and the places through which they travelled, Cawnpore and the events that had taken place there represented the first place and the first conflict that most British women described on their journey away from Lucknow. Julia Inglis wrote, for example,

> My feelings on entering Cawnpore were indeed most painful. The moon was bright, and revealed to us the sad spectacle of ruined houses, trees cut down, or branches stripped off, everything reminding us of the horrors that had been enacted in the place, and making us feel thoroughly miserable.
>
> (Inglis 1892, 214)

The 'horrors' to which Inglis refers had taken place in June, and Julia Inglis and other British women had learned about them from the first 'relief' force that reached Lucknow in September. As Russell wrote in 1858, 'the peculiar aggravation of the Cawnpore massacres was this, that the deed was done by a subject race – by black men who dared to shed the blood of their masters, and that of poor helpless ladies and children' (29), and the sensationalist objectification of British women in the 'mutiny' reached its lurid peak in contemporary and historical representations of these events. In June 1857, the British population at Cawnpore lived under siege in entrenchments for three weeks. After accepting terms of release, most were killed on 27 June. The surviving 210 women and their children were kept as prisoners until 15 July when, because of the approach of the British army – the same soldiers who were to reach Lucknow in September – they were also killed.[26]

In their diary entries in September 1857, British women at Lucknow recorded their responses to events at Cawnpore in ways that echoed representations in newspapers and subsequent histories, by focusing not only on the deaths of British women and children but also on the place where they had died. As Julia Inglis wrote, 'I believe in the annals of history no records could be found of a deed of such unexampled atrocity as murdering in cold blood so many defenceless women and children. My very heart sickens at the thought' (1858, 23), while Maria Germon wrote that 'They say the place where the murders took place was a horrible sight – not a soldier left it with a dry eye' (1957, 98–9). The violent consequences of British soldiers travelling through Cawnpore and visiting the place where British women and children died have been well documented. This place was the site of some of the most brutal retribution against Indians in the 'mutiny',[27] and eye-witness accounts and drawings that were published in

Anglo-Indian and British newspapers continued to fuel popular demands for vengeance against the deaths of British women (Sharpe 1993).[28] However, little attention has been paid to the accounts written by British women who visited Cawnpore on their journey away from Lucknow.

The British women evacuated from Lucknow stayed in Cawnpore until the night of 3 December, staying either in tents or in the artillery barracks. While there, Colina Brydon, Adelaide Case, Katherine Harris and Julia Inglis visited the entrenchments where the British population had lived for three weeks under siege (Figure 3), but were unable to visit the house where British women and children were later killed as it had by now been recaptured by Indian forces. Their visit to the entrenchments, however, marked a turning point in their writings about leaving Lucknow and played an important part in reconstituting their imperial subjectivity. Unlike their diary entries in September while they were living under siege and fearing the same fate suffered by British women at Cawnpore, their accounts about visiting the entrenchments reflect a simultaneous recognition not only of the extent of the imperial crisis but also of their own place as survivors. As Adelaide Case wrote,

> I could not have believed, had I not seen it, that their abode had been so wretched. . . . The intrenchment is scarcely even a good-sized ditch, and yet, at times, it was safer for the poor ladies to take their chairs and sit there than to remain in the miserable building which scarcely afforded a shelter from the sun.
>
> (Case 1858, 314)

Figure 3 'The Entrenchment at Cawnpore', *Illustrated London News*, 24 October 1857.

Julia Inglis also commented on the paucity of defences, but, unlike Case, found that seeing such a sight made it no easier to comprehend, writing that 'As I looked, I thought how small were the troubles and trials of Lucknow in comparison. The agony and miseries these poor creatures must have suffered defies even imagination to conceive' (1892, 218).

Visiting Cawnpore marked a turning point for British women travelling away from Lucknow. Once they received letters from family and friends at Dilkusha Palace, they learned how their situation had been represented at 'home', and they began to describe their place at the heart of an imperial crisis. However, it was only once they saw the destruction of the city and the entrenchments at Cawnpore that they began to represent both the conflict and the places through which they travelled in less detached ways. In particular, touring the entrenchments caused them not only to reflect on the fate of other British women in the 'mutiny', but also, for the first time, to represent themselves as survivors, travelling away from the confinement of both Lucknow and Cawnpore. Katherine Harris, for example, kept a page of a Bible as a 'relic' of touring the entrenchments at Cawnpore and, the next day, wrote that 'We breathe more freely now we are out of that mournful place, Cawnpore; and as the whole road between it and Allahabad is lined with our troops coming up country, I trust it is pretty safe' (1858, 183–4).

RECONSTRUCTING HOME

The British women evacuated from Lucknow reached the safety of Allahabad on 7 December, travelling the last forty miles by train. As Katherine Harris wrote, 'it seemed delightfully home-like and natural to be once more on a railroad' (1858, 187), and, after 'an almost overpowering' welcome from soldiers at the station,[29] the British women all wrote about living in Allahabad in terms of settled and familiar domesticity. Most of the diarists stayed at Allahabad for six weeks until they could travel by steamer to Calcutta and, during that time, lived in tents belonging to the Governor General that were erected in the grounds of the fort. In marked contrast to their descriptions of Cawnpore, British women described their temporary home in picturesque terms, with Katherine Harris writing, for example, that 'our camp is really very pretty, beautifully pitched in a square, with the large dining tent in the centre, on a lovely piece of turf, with trees all around' (1858, 189). The tents themselves were described as spacious and luxurious, affording privacy for the first time since the siege of Lucknow had begun. For Maria Germon, 'It was a great luxury to be quite by oneself after the many months we had been herded together' (1957, 134), while Katherine Harris shared a room with only her husband for the first time since 24 May (1858).

Although Francis Wells wrote to her father that 'we are very impatient at being detained here so long',[30] most British women wrote more favourably about the prospect of spending time in a peaceful place. As Katherine Harris wrote, 'The luxurious feeling of rest and peace and safety here is perfectly indescribable; one can scarcely realise it or know what to make of it after the excitement, anxiety, and turmoil of the last six months' (1858, 189).

Living in the security of Allahabad led many women to reflect not only on the alienation and dangers of their lives under siege but also on their desire to be at home again. While visiting Cawnpore had revealed to many women the extent of the imperial crisis and their place within it, living in Allahabad enabled such women to begin to reconstruct their domestic as well as their imperial subjectivity. This domestic and imperial reconstruction took different forms, such as entertaining guests for meals and attending church and, for the first time since leaving Lucknow, class differences among British women became increasingly evident. Any common experience as white, British women living under siege were lost in the face of class difference. Katherine Harris wrote, for example, that 'It is wonderful how little that class of people seem to feel things that would almost kill a lady' (1858, 56) and, on the desertion of many of their Indian servants, soldiers' wives were often employed as cooks and nurses. However, soldiers' wives remained largely invisible and unrecorded in diaries both about the siege and about leaving Lucknow. It was only once they were living in safety at Allahabad, with enough Indian servants to re-establish an imperial hierarchy on a domestic scale, that the diarists began not only to represent but also to reconstitute their class position in relation to other British women. The class differences between women were spatially as well as socially inscribed – the luxurious tents of the 'ladies' contrasting with the crowded barracks where the wives and widows of soldiers lived. Three days after helping to distribute 'shoes, stockings, pocket handkerchiefs, combs, and hair brushes' to the 'ladies' of Lucknow,[31] Katherine Harris toured the barracks and collected names of soldiers' widows who would each receive a black dress, noting that 'Some of the poor things are in great distress, having come out of Lucknow with only the clothes they wore' (Harris 1858, 189, 192).

In Allahabad, the time and security enabled familiar routines and class relations to be re-established. In the hierarchical structure of an army regiment, this was most clearly apparent in the contact between Julia Inglis and the wives and widows of the 32nd Regiment.[32] On Christmas Day, as well as entertaining four people for dinner, Inglis wrote that 'I also gave the women and children who were left of the 32nd a dinner. It was anything but a festive sight to me. There were now only seventeen women, and nearly all were widows, and every child present had lost one or both parents' (1892, 224). Katherine Harris had helped to set up a school for these and other children whom, when she visited the barracks, she had found 'running about wild' (1858, 192), and Julia Inglis asked to teach the children from the 32nd Regiment on Sundays. As Inglis wrote,

> The first time was very trying, as the remembrance of all that had happened since I last saw them, and the thoughts of their companions who had died so terrible a death, quite overcame all the children, and it was some time before I could continue speaking to them.[33]

(Inglis 1892, 224)

This suggests that Julia Inglis was seeing the children from the 32nd Regiment for the first time since the beginning of the siege. Not only did the security of Allahabad enable British women diarists to represent their domestic as well as imperial subjectivity, but it also allowed them to reconstitute their domestic subjectivity in class-specific ways. Most notably, writers such as Katherine Harris and Julia Inglis wrote about their contact with the wives, widows and children of British soldiers in terms of visits, hospitality and teaching, with such contact helping to reconstruct their own class as well as gendered identity.

Other women, however, found it harder to reconstruct their domestic subjectivity while living at Allahabad. Their temporary home led several British women to write about their memories and desire for their own homes, either in Britain or India. Francis Wells, for example, imagined home in Britain in picturesque terms, writing to her father, 'How I shall enjoy home after all this, the sweet roses and birds and above all the peace and quiet.'[34] In contrast, sitting in the garden of the house where she was staying in cantonments led Katherine Bartrum to miss her lost home in India:

> The scene around is very pretty, but it brings so vividly before me our dear little bungalow at Gonda that it makes my heart very sad in thinking of the days that are no more. We wander through the deserted houses around, most of which have been burnt to the ground. . . . The scene of desolation which the place presents is very sad: so many happy homes having been utterly destroyed in the past year.
>
> (Bartrum 1858, 61)

Living in the security of Allahabad, away from the dangers of Lucknow and Cawnpore, enabled British women such as Katherine Harris and Julia Inglis to reconstitute their domestic as well as their imperial subjectivity. However, for women such as Katherine Bartrum, whose husband had been killed at Lucknow and who was travelling away from her home in India for the last time, both her domestic and imperial subjectivities seem irretrievably lost.

The first steamer left Allahabad shortly before Christmas with 'all the widows and sick ladies' on board (Inglis 1892, 223). Most of the British women from Lucknow began the three-week journey to Calcutta in mid-January, and described their progress as slow, crowded and, at times, hazardous. Katherine Bartrum travelled through familiar places on her last journey away from her Indian home, writing about Dinapore, for example, that 'Every spot here reminds me of bygone days, when we were stationed at this place for seven months, when everything shone so brightly around me. Here my darling child was born. Saw Mr Burge, the chaplain, but most of those whom we knew here have returned home' (1858, 63). Other British women, however, wrote little about their journey down the Ganges and their memories of past, happy homes in India. Instead, most British women focused on their arrival in Calcutta and on the domestic and imperial security of the last place where they stayed in India before travelling 'home' to Britain. Although subsequent steamers 'were suffered to land quietly'

(Inglis 1892, 227), the first steamer to arrive had been welcomed with Royal salutes, a red carpet, an official welcoming party, and a crowd of onlookers on the quay, reflecting 'the deep but cordial sympathy of the whole European population'.[35] The Lucknow 'heroines' were represented in person for the first time by Anglo-Indian and British newspapers, which reported, for example,

> The black dresses of most of the ladies told the tale of their bereavement, whilst the pallid faces, the downcast looks, and the slow walk, bore evidence of the great sufferings they must have undergone both in mind and body. The solemn procession thus passed on, and was handed into carriages which conveyed them to their temporary home.
>
> (*Lady's Newspaper* and *Pictorial Times*, 20 February 1858)

Many British women found their temporary homes in Calcutta with friends or relations or in houses provided by the Lucknow Relief Committee, which had been set up in Calcutta and was supported by funds raised in Britain. Ann Ellen Huxham stayed with her sister in her 'beautiful home' (Huxham n.d.) and Julia Inglis wrote that, 'Finding myself once again in a comfortable house, with all the appurtenances of civilization around me, made me feel quite strange' (1892, 227).[36] While letters from 'home' that the British women received at Dilkusha Palace had suggested the extent of the imperial crisis for the first time, contact with 'home' through the Lucknow Relief Committee helped British women to reconstitute their domestic and imperial subjectivity. The Lucknow Relief Committee was a practical response to the crisis in India. Its funds enabled women like Maria Germon to stay in 'luxurious style' and also contributed towards clothing, a passage back to Britain and some compensation for lost goods and property. In this way, the domestic and imperial security of Calcutta, epitomized by the comfort of the 'temporary homes' in which British women stayed, was directly supported by the British public. As Katherine Bartrum wrote, for example,

> I can never feel sufficiently grateful for the universal kindness and sympathy I received during my short stay in Calcutta. The hand of friendship was held out even by strangers, and everything was done to alleviate our sorrow and distress. Most nobly did England respond to the cry of the widow and the orphan.
>
> (Bartrum 1885, 65)

Travelling and writing home from Lucknow, British women diarists came not only to represent but also to reconstitute their domestic and imperial subjectivities on different scales. Once they had arrived in Calcutta, and stayed in temporary homes with friends, relatives, or provided by the Lucknow Relief Committee, such women could begin to reconstruct their ideas of home, not only in Calcutta but also in Britain.

CONCLUSIONS

In this chapter, I have focused on the writings of several British women as they travelled from Lucknow to Calcutta between November 1857 and February 1858. In their letters and diaries, British women not only represented but also began to reconstitute their domestic and imperial subjectivity as they travelled away from the dangers of Lucknow and Cawnpore towards the safety of Allahabad and Calcutta. Imaginative geographies of home and away existed on different scales and helped to reconstitute the domestic and imperial subjectivity of British women travelling and writing home from Lucknow. Throughout their accounts, British women represented the imperial crisis in domestic terms, focusing on everyday life and the desertion of their servants during the siege. In the first few days after leaving Lucknow, they continued to represent their domestic concerns about what to wear and pack and seemed in many cases to be detached not only from the imperial conflict but also from the places through which they passed. It was only once they received letters from 'home' at Dilkusha Palace that the British women evacuated from Lucknow began to realize the extent of the imperial crisis. Once they reached Cawnpore, there was a simultaneous recognition not only of where fighting had taken place but also of their own role as survivors of the conflict. Soon afterwards, arriving in the security of Allahabad, many women could begin to reconstruct their domestic as well as their imperial subjectivity. This domestic reconstruction was class as well as gender specific, shown by the social and spatial distance between 'ladies' and soldiers' wives and widows. Such a process of reconstruction continued when the British women arrived in Calcutta and lived in temporary homes financed from Britain. Here, the activities of the Lucknow Relief Committee represented the domestic management of an imperial crisis, facilitating and supporting both the domestic and imperial security and subjectivity of the British women.

The writings of British women leaving Lucknow represent their experiences of collective and forced rather than individual and voluntary travel. As such, their writings differ from the accounts of individual women travellers who were able to transgress in social as well as spatial terms the confines of 'home' in the context of imperialism. The letters and diaries from Lucknow rather reveal the reconstruction rather than transgression of their domestic subjectivity at a time of imperial crisis. Just as the spaces of home and away were more blurred than distinct through imperial travel, so too were the domestic and imperial subjectivities of British women travelling and writing home from Lucknow. The imaginative geographies of home and away that helped to reconstruct their domestic and imperial subjectivities existed not only within India but also between Britain and India. British women played often ambivalent roles as domestic and imperial subjects travelling and living in India, as shown by representations of their evacuation from Lucknow and by a bitter debate about the place of British women in India in the years following the 'mutiny'.

One consequence of the 'mutiny' for imperial India was the Royal Proclamation of 1858, and subsequent Government of India Acts, which replaced the rule of the East India Company with that of the British Crown. Another consequence of the 'mutiny' was that an increasing number of British women came to live in India. However, the very presence of British women in India, and their domestic, home-making responsibilities, were represented in ambivalent terms. Some commentators viewed British women as essential, not only to reproduce legitimate imperial rulers but also to reproduce the social, moral and domestic values legitimating imperial rule. For example, the authors of a best-selling guide to housekeeping in India described the home as 'that unit of civilisation where father and children, master and servant, employer and employed, can learn their several duties' (Steel and Gardiner 1907, 7). Other commentators, however, argued that the presence of British women in India undermined British rule by leading to greater racial segregation. It was argued that increasing numbers of British women living in India over the course of the nineteenth century had helped to create a separate sphere of exclusively British domestic, social and moral life. As Wilfred Scawen Blunt wrote in 1885,

> the Englishwoman in India during the last thirty years has been the cause of half the bitter feelings between race and race. It was her presence at Cawnpore and Lucknow that pointed the sword of revenge after the Mutiny, and it is her constantly increasing influence now that widens the gulf of ill-feeling and makes amalgamation daily more impossible.
>
> (Blunt 1885, 47)

Representations both of and by British women travelling away from Lucknow centred on their often ambivalent place as both domestic and imperial subjects. In the years following the 'mutiny', bitter debates about the place of British women in India also centred on their domestic and imperial subjectivity. As I have shown, British women travelling and writing home from Lucknow not only represented but also reconstructed their domestic and imperial subjectivity. Their accounts reveal connections between domesticity and imperialism and also between the imaginative geographies of home and away that underpinned – and yet potentially undermined – both imperial rule and the travel on which it relied.

NOTES

1 Following Thomas Metcalf, the events of 1857–8 were 'something more than a sepoy mutiny, but something less than a national revolt' (Metcalf 1965, 60). While the unrest largely took the form of an uprising by Indian soldiers against their British officers, in some places, particularly the recently annexed province of Oudh, it also took the form of widespread agrarian unrest. The uprising was brutally suppressed by several thousand troops sent from Britain by June 1858. See Hibbert (1978).
2 The *Illustrated London News*, for example, described 'the wholesale butcheries of Englishmen and women, and the foul indignities previously perpetrated upon the latter' (29 August 1857) and asked, 'what do those who cry out for mercy to such

wretches say of the murder of helpless babies and unoffending women? and of the almost incredible indignities and cruelties committed upon English ladies – cruelties so horrible that their mere mention is almost an offence in itself?' (8 August 1857). For further discussion of the sensationalist objectification of British women in the 'mutiny', see Sharpe (1991) and Sharpe (1993), where she argues that 'A discourse on rape . . . helped manage the crisis in authority so crucial to colonial self-representation at the time' (p. 67).

3 The Residency compound consisted of 33 acres of land around the Residency building, which included 'a large number of bungalows, houses, small palaces, and fortified gates' (Edwardes 1957, 17).

4 Estimates of the numbers under siege at Lucknow vary. Innes states that there were 3,000 people under siege, of whom 1,392 were Indian and 1,608 were British and others of European descent. Innes also estimates that there were 1,720 combatants and 1,280 non-combatants.

5 Katherine Bartrum (1858), Adelaide Case (1858), Katherine Harris (1858), Ann Ellen Huxham (n.d.) and Julia Inglis (1892) published their diaries about Lucknow and Inglis also published a letter for private circulation. Unpublished accounts include a diary kept by Fanny Boileau (Boileau Papers, Centre of South Asian Studies, University of Cambridge: extracts from the diary of Fanny Elizabeth Boileau, Lucknow, 1857) and letters written by Francis Wells to her father (Bernars Papers, Centre of South Asian Studies, University of Cambridge). Diaries written by Maria Germon and Colina Brydon were unpublished in their lifetimes, but subsequently published for the first time in 1957 (Germon) and 1978 (Brydon).

6 Also see Grewal (1996) and Marangoly George (1994) for more discussion about the connections between domesticity and imperialism.

7 *Illustrated London News*, not dated, from The Mutiny Scrap Books, School of Oriental and African Studies, London, MS 380484.

8 Arthur Dashwood, who was born during the siege, wrote that annual commemorative dinners continued until 1913. A. F. Dashwood, 'Untimely Arrival at the Siege of Lucknow', *The Listener*, 2 December 1936, reprinted in Brydon (1978).

9 *The Solomon Family of Painters*, Exhibition Catalogue, The Geffrye Museum, London, 1985, p. 56. Thanks to Adrienne Avery-Gray at The Museum and Art Gallery of Leicester and Louise West at The Geffrye Museum for their help.

10 Events at Cawnpore will be discussed below.

11 General order from Sir Colin Campbell, 21 November 1857, quoted in the *Lady's Newspaper and Pictorial Times*, 16 January 1858. In light of powerful discourses of masculine honour, heroism and revenge, the prestige of the British army was inextricably linked to its ability to save and protect British women. As an example of such discourses, the *Illustrated London News* wrote that 'We hear with pain, but not perhaps with horror, of the deaths of our brave officers and soldiers slain by the mutineers, for it is the soldier's business to confront death in all its shapes; but when we read of the atrocities committed upon our women and children the heart of England is stirred; and the sorrow for their fate, great as it is, is overshadowed by the execration which we feel for their unmanly assassins, and by the grim determination that Justice, full and unwavering, shall be done upon them'. (*Illustrated London News*, 5 September 1857).

12 Quoted in the *Lady's Newspaper and Pictorial Times*, 16 January 1858.

13 Maria Germon was married to Captain Richard Charles Germon of the 13th Bengal Native Infantry.

14 Katherine Harris was married to one of the two chaplains at the siege of Lucknow.

15 Katherine Bartrum was the widow of a doctor in the Bengal Medical Service who died on 26 September 1857, in the first 'relief' of Lucknow. Adelaide Case was the

widow of Colonel Case of the 32nd Regiment who was killed in the unsuccessful battle of Chinhut in June 1857, which marked the beginning of the siege of the Lucknow Residency. She kept a diary 'for the perusal of my relatives in England, and with no view whatever to publication' (Case 1858, iii). Ann Ellen Huxham – wife of a captain in the Indian Army – published *A Personal Narrative of the Siege of Lucknow* as an undated pamphlet. Although she published a letter to her mother for private circulation in 1858, Julia Inglis did not publish her diary until 1892. She then did so because, in her opinion, 'a thoroughly clear and accurate account [of the siege of Lucknow] has not been given' (Inglis, 1892, v).

16 Katherine Harris, for example, wrote that 'As no lady's diary has hitherto been given to the public, the friends of the writer have thought that it might interest others, beyond the family circle, to communicate additional information on a subject in which the British nation feels so deep an interest'(1858, iii).

17 As she continued, she also published her diary 'to show how wonderfully she was protected in perils and dangers of no ordinary kind, and how, when called to drink deeply of the cup of human sorrow, the arm of the Lord was her stay, a "rock of defence in the day of trouble"'.

18 As Jenny Sharpe writes, 'Bloodshed and massacre, the common tropes for thinking about 1857, overshadow the less dramatic effect the rebellion had on everyday life. English women's writings are especially valuable for their descriptions of the domestic disorder in a colonial world turned upside down' (Sharpe 1993, 62).

19 Also see Blake (1990).

20 For more on the work of soldiers' wives and the class divide between the wives of soldiers and officers, see Trustram (1984).

21 Inglis described herself as 'thunderstruck' (Inglis 1892, 197) and Germon described the news as 'like a thunderbolt' (Germon 1957, 119).

22 As Germon continues, she mounted her pony 'with great difficulty', assisted by her husband and Captain Weston, who, with a large party, 'were in fits of laughter' (121). Once she had left the Residency, she had to dismount at one point to lead her pony through a trench and needed the assistance of three men and a chair to remount. Because of these difficulties, she chose to ride rather than walk through other dangerous parts of the route.

23 Sikandar Bagh was a two-storey house in a large, walled garden, built by Nawab Wajid Ali Shah (Taylor 1994).

24 As Ann Ellen Huxham wrote, 'We felt rather forlorn until our attention was directed to a splendid feast, spread under some trees, consisting of cold beef and chickens, and bread and butter, luxuries to which we had been strangers for many months. I don't think we ever enjoyed any meal in our lives so much as that *al fresco* supper' (Huxham, n.d.).

25 The Dilkusha Palace was built as a hunting lodge by Nawab Saadat Ali Khan and stood in an extensive deer park (Taylor 1994). Letters were dated up to 18 September and Francis Wells, for example, received nineteen (F. Wells to her father, Dr Fox, from Allahabad, 12 December 1857).

26 Nineteenth-century accounts of events at Cawnpore include Shepherd (1894), Thomson (1859) and Trevelyan (1886). For a more recent debate, see Mukherjee (1990), English (1994) and Mukherjee's reply (1994).

27 See Hibbert (1978) for discussion of the punishments ordered by General Neill.

28 The *Bengal Hurkaru*, for example, published an engraving of 'the interior of the house at Cawnpore where ladies and children were massacred' on 25 November 1857, with a caption saying that 'The floors were slippery with blood and the walls daubed with it. The courtyard was soaking, and in dragging the bodies across it the sand and blood had formed a sort of red paste'. Letters describing the scene were published in the *Illustrated London News* on 19 September, 26 September and 10 October.

29 Bartrum, Germon and Inglis all described their welcome at Allahabad in this way.
30 Francis Wells to her father, Dr Fox, from Allahabad, 2 January 1858.
31 These items had been sent from Calcutta by Lady Canning, the wife of the Governor General.
32 See Trustram (1984) for more on the social gulf between the wives of officers and soldiers.
33 In May 1857, before she knew Julia Inglis personally, Katherine Harris had written in her diary about the regimental school of the 32nd Regiment, noting that 'Mrs Inglis took great pains with it, and had the children at her own house on Sundays. It is such a rare thing in this country to find ladies interesting themselves about the poor women and children; but the Inglises, from what I hear of them, must be excellent people' (Harris 1858, 16–17).
34 Letter from Francis Wells to her father, Dr Fox, from Allahabad, 2 January 1858.
35 *Lady's Newspaper and Pictorial Times*, 20 February 1858.
36 Julia Inglis stayed with Sir Charles Jackson, the chief justice, and his wife.

REFERENCES

Bartrum, K. (1858) *A Widow's Reminiscences of the Siege of Lucknow*, London: James Nesbit and Co.

Blake, S.L. (1990) 'A woman's trek: what difference does gender make?', *Women's Studies International Forum* 13(4), pp. 347–55.

Blunt, A. (1994) *Travel, Gender and Imperialism: Mary Kingsley and West Africa*, New York: Guilford.

Blunt, A. and Rose, G. (eds) (1994) *Writing Women and Space: Colonial and postcolonial geographies*, New York: Guilford.

Blunt, W.S. (1885) *Ideas about India*, London: Kegan Paul, Trench and Co.

Brydon, C. (1978) *The Lucknow Siege Diary of Mrs C.M. Brydon*, ed. and published by C. de L.W. fforde.

Case, A. (1858) *Day by Day at Lucknow: A journal of the siege of Lucknow*, London: Richard Bentley.

Edwardes, M. (1957) 'Introduction' to M. Germon, *Journal of the Siege of Lucknow: An episode of the Indian Mutiny*, ed. M. Edwardes, London: Constable.

Engels, F. (1858) 'The capture of Lucknow', New York *Daily Tribune*, 30 April 1858. Reproduced in K. Marx and F. Engels (1959) *The First Indian War of Independence 1857–1859*, Moscow: Progress Publishers, pp. 122–3.

English, B. (1994) 'Debate: the Kanpur Massacres in India in the revolt of 1857', *Past and Present* 142, pp. 169–78.

Germon, M. (1957) *Journal of the Siege of Lucknow: An episode of the Indian Mutiny*, ed. M. Edwardes, London: Constable.

Grewal, I. (1996) *Home and Harem: Nation, gender, empire and the cultures of travel*, Durham, NC: Duke University Press.

Harrington, P. (1993) *British Artists and War: The face of battle in paintings and prints, 1700–1914*, London: Greenhill Books.

Harris, K. (1858) *A Lady's Diary of the Siege of Lucknow*, London: John Murray.

Hibbert, C. (1978) *The Great Mutiny: India 1857*, London: Penguin.

Huxham, A.E. (n.d.) *A Personal Narrative of the Siege of Lucknow*, pamphlet.

Inglis, J. (1858) *Letters containing extracts from a Journal Kept by Mrs Julia Inglis during the siege of Lucknow*, London: privately printed.

Inglis, J. (1892) *The Siege of Lucknow, A Diary*, London: James R. Osgood, McIlvaine and Co.

Innes, M. (1895) *Lucknow and Oude in the Mutiny: A narrative and a study*, London: A.D. Innes and Co.

Kaye, J. (1876) *A History of the Sepoy War in India 1857–8*, vol. 2, London: W.H. Allen.

McClintock, A. (1995) *Imperial Leather: Race, gender and sexuality in the colonial contest*, New York: Routledge.

Marangoly George, R. (1994) 'Homes in the empire, empires in the home', *Cultural Critique* 15, pp. 95–127.

Metcalf, T. (1965) *The Aftermath of Revolt: India 1857–1870*, Princeton: Princeton University Press.

Mills, S. (1991) *Discourses of Difference: An analysis of women's travel writing and colonialism*, London: Routledge.

—— (1996) 'Gender and colonial space', *Gender, Place and Culture* 3(2), pp. 125–48.

Mukherjee, R. (1990) ' "Satan let loose upon the earth": the Kanpur Massacres in India in the revolt of 1857', *Past and Present* 128, pp. 92–116.

—— (1994) 'Reply to "Debate: the Kanpur Massacres in India in the revolt of 1857" ', *Past and Present* 142, pp. 178–89.

Russell, W.H. (1957) *My Indian Mutiny Diary*, ed. M. Edwardes, London: Cassell.

Said, E. (1979) *Orientalism*, New York: Vintage.

Sharpe, J. (1991) 'The unspeakable limits of rape: colonial violence and counter-insurgency', *Genders* 10, pp. 25–46.

—— (1993) *Allegories of Empire: The figure of woman in the colonial text*, Minneapolis: University of Minnesota Press.

Shepherd, J. (1894) *A Personal Narrative of the Outbreak and Massacre at Cawnpore, During the Sepoy Revolt of 1857*, 4th edition, Lucknow: Methodist Publishing House.

Steel, F.A. and Gardiner, G. (1907) *The Complete Indian Housekeeper and Cook*, 5th edition, London: William Heinemann.

Stokes, E. (1978) *The Peasant and the Raj: Studies in agrarian society and peasant rebellion in colonial India*, Cambridge: Cambridge University Press.

Taylor, P.J.O. (1994) 'Empire and the pursuit of aesthetics', *The Taj Magazine: Lucknow* 23(1), pp. 42–53.

Thomson, M. (1859) *The Story of Cawnpore*, London: Richard Bentley.

Trevelyan, G. (1886) *Cawnpore*, 2nd edition, London: Macmillan.

Trustram, M. (1984) *Women of the Regiment: Marriage and the Victorian army*, Cambridge: Cambridge University Press.

6 Scripting Egypt

Orientalism and the cultures of travel

Derek Gregory

The Gods seem to have arranged the Nile Valley sights so that the traveller can read Baedeker's illuminating description of the next place on the programme in plenty of time to appear intelligent and profit by the visit, and also to appreciate the joyous donkey ride to and from some grave or shrine, without even hurrying over a meal.

Wilfrid Thomason Grenfell

TRAVEL WRITING

In 1845 W.H. Bartlett was hesitant to contribute to the growing library of books on Egypt. 'To add another book on Egypt to the number that have already appeared', he wrote, 'may almost appear like a piece of presumption.' But he distinguished between the 'army' of erudite *savants* schooled in archaeology, history and natural history – a reference to the scholars who had accompanied Napoleon's army of occupation in Egypt between 1798 and 1801 – and those who, like himself, were enlisted in what he called the 'flying corps of light-armed skirmishers, who, going lightly over the ground, busy themselves chiefly with its picturesque aspect' and 'aim at giving lively impressions of actual sights' (Bartlett 1849, iii). Whatever the merits of the distinction, there was no doubt about Bartlett's success: his account, *The Nile Boat, or glimpses of the land of Egypt*, turned out to be one of the canonical texts of travel in Egypt.

Succeeding authors made the same show of reluctance and then, just like Bartlett, pressed on regardless. Thirty-odd years later, when Charles Warner set out to record his impressions, he observed 'that if the lines written about Egypt were laid over the country, every part of it would be covered by as many as three hundred and sixty-five lines to the inch' (Warner 1876, vi). The imagery was irresistible. Charles Leland advised travellers that 'for the practical part of your journey you consult guide-books and all kinds of literary Nilometers to see how high it will rise in prices or how low it will ebb in your purse' (Leland 1875, 264). And T.G. Appleton conceded that, just as 'every year a little deposit of useful mud is left by the Nile upon its banks, [so] every year sees deposited upon

counters of the London booksellers the turbid overflow of journalizing travel'
(Appleton 1876, i).

By the end of the nineteenth century, it seemed as if virtually every European
or American traveller to Egypt had felt compelled to write about the experience.
In this essay I seek to prise open that connective imperative between 'travel' and
'writing'. My argument is triangulated by three ideas: the construction of the
Orient as a *theatre*; the representation of other places and landscapes as a *text*; and
the production of travel and tourism as a *scripting*. Edward Said's critique of
Orientalism has made the first two commonplaces, but I need to say something
about them in order to bring the third – which is my main concern – into focus.

THEATRE, TEXT

Many travel writers described Egypt as a theatre. Said claimed that the very idea of
representation is a theatrical one, and that within the discourses of Orientalism
'the East' was typically constructed as 'a closed field, a theatrical stage affixed to
Europe' (Said 1979, 63). In Egypt travel writers invested the metaphor in a visual
economy that implied not only spectacle but also illusion, even the not-quite-real.
'One had to rub one's eyes', declared Harriet Martineau, who travelled up the
Nile in 1848, 'to be sure that one was not in a theatre. . . . [It] was like a subli-
mated opera scene' (Martineau 1848, 69). At the limit this magic-theatre turned
into a house of distortion and deception. When Warner described the streets of
Cairo as an elaborate masquerade, for example, he admitted to the uneasy feeling
that 'there is a mask of duplicity and concealment behind which the Orientals live'
(1876, 48). Orientalism's counter-strategy was to construct Egypt as a *transparent* space, exposed to the gaze of the observer who had the power – and the
duty – to sweep aside the mask or, in a visible sexualized project of dis-covery, to
remove the veil.

Travel writers also described Egypt as a text or series of texts. 'This country is a
palimpsest', wrote Lady Gordon in 1863, 'in which the Bible is written over
Herodotus, and the Koran over that. In the towns the Koran is most visible, in the
country Herodotus' (Gordon 1969, 65). Ten years later Laurent Laporte
deployed an even older figure to describe Egypt as 'a hieroglyphic that the whole
world is trying to decipher' (Laporte 1872, 96) and in 1873 Amelia Edwards
described Egypt as 'a Great Book', though she confessed that it was 'not very easy
reading' (Edwards 1877, 70). Orientalism's counter-strategy was to construct
Egypt as a *legible* space whose cultural inscriptions, however faint or obscure,
could be deciphered by the educated reader: travel thus became an intrinsically
hermeneutic project.

The production of Egypt as a transparent and legible space was part of the
apparatus of what Timothy Mitchell calls 'the world-as-exhibition'. In his view,
colonial power 'required the country to become readable, like a book, in our own
sense of such a term' (Mitchell 1988, 33). One might say much the same of
the theatrical metaphor: it too staged Egypt as an enframed exhibit set up before a

privileged audience. Metaphors of 'theatre' and 'text' are not innocent, therefore, and speak to power and the production of spaces and to the practices that take place within (and indeed through) them. But these twin strategies do not coincide with any simple distinction between visual and textual forms; they find common ground in what Said calls 'the textual attitude' which he thought was exemplified by the kind of travel writings I consider in this essay. From this perspective, he wrote, 'people, places and experiences can always be described by a book, so much so that the book (or text) acquires a greater authority, and use, even than the actuality it describes' (Said 1979, 93). I want to retain many of these implications, but to work with a different metaphor (although it evidently trades on the other two): what I call a 'scripting'.

SCRIPTING, TRAVELLING AND WRITING

To describe the cultural practices involved in travel and tourism in these terms is not original. In his study of tourism in Europe during the long nineteenth century, for example, James Buzard writes of 'the scripted continent'; but he does so in ways that constantly fold travel back into the text. Although he acknowledges that the postwar tour was regulated by a series of 'guiding texts', much more so than the Grand Tour had been during the eighteenth century, he is plainly more interested in the presence of those texts in other texts. Thus the aspiring travel writer 'had to work within the boundaries mapped out by those prior texts or somehow to stake out new territories within one's own text' (Buzard 1993, 156). All this may be granted, but the spaces that Buzard invokes – 'territories' marked by 'boundaries' – are conspicuously textual and do not directly engage with the spaces through which writers travelled.

I have chosen to speak of a 'scripting' precisely because it accentuates the production (and consumption) of spaces that reach beyond the narrowly textual, and also because it foregrounds the performative and so brings into view practices that take place on the ground. Accordingly, I understand a 'scripting' as a developing series of steps and signals, part structured and part improvised, that produces a narrativized sequence of interactions through which roles are made and remade by soliciting responses and responding to cues.[1] I don't mean this to reduce travel and tourism to the dramaturgical, but thinking about travel writing in this way has several important implications.

In the first place, it directs our attention to the ways in which travel writing is intimately involved in the 'staging' of particular places: in the simultaneous production of 'sites' that are linked in a time–space itinerary and 'sights' that are organized into a hierarchy of cultural significance. Travel scripting produces a serialized space of constructed visibility that allows and sometimes even requires specific objects to be seen in specific ways by a specific audience. Places are thus signposted so that tourists can find them as 'sites' and locate them within an imaginative landscape where they become meaningful as 'sights'. In the second place, scripting reminds us that travel writings are literally 'passages' that mark and are marked by worlds in motion. They are always more (and always less) than a direct record of experience, but they all carry within them traces of the physical movement of embodied subjects through material landscapes. Their reading thus requires us to move beyond the spaces of the desk and the library. In the third place, scripting indicates that cultures of travel are collective and contradictory. The routes of most tourists are routinized, and each trip in its turn contributes to the layering and sedimentation of powerful imaginative geographies that shape (though they do not fully determine) the expectations and experiences of subsequent travellers. But these touring productions also have multiple authors in quite another sense, because they depend upon – and can be interrupted, dislocated and reworked by – the practices of local people who are involved in them.

In what follows I use these ideas to consider the ways in which Egypt was scripted by European and American travel writers between 1820 and 1920. I focus on scriptings that incorporated textual practices – reading and writing – as central moments in their production and reproduction, and on the ways in which they were drawn into the performances of travel and tourism on and off the printed page.

SITES AND SIGHTS

When George Curtis described how he sailed 'down into the heart of Egypt and into the remote past, living in fact by books and by eye-sight' (Curtis 1858, 154) he was pointing out the intimate connections between reading and sightseeing: the conversion of 'sites' into 'sights' was mediated in all sorts of ways by travel writing. Travellers prepared for their journey by studying in advance, and they also took large numbers of books with them. In the middle decades of the nineteenth century those who hired a *dahabeeah* – a large house-boat with cross-sails – soon discovered that books were, as William Prime put it, 'an essential to the pleasures of the voyage' (Prime 1857, 495). But they were scarcely light reading. 'The volumes selected for the library of the dahabeah', intoned the Reverend Smith, 'will be principally confined to those which treat of Egyptology as well as of those more distant countries whence the Nile takes its rise; in all of which, and the speculations and discussions thereupon, the travelling Hawager [foreigner] will soon be deeply absorbed' (Smith 1868a, xvii–xviii). He was not unusual in recommending some forty titles. It was no different on the tourist steamer later in the century: 'You are lost in Egypt if you do not read up,' warned an unusually sober Douglas Sladen. 'The only fault I have to find with the cabins is that the electric light goes out at eleven, for on the Nile, more than anywhere else, you want to do a good deal of reading at night. All the best books on Egypt are in the ship's library' (Sladen 1908, 414).

The essential volumes included not only classical texts like Herodotus but also modern works of scholarship like Wilkinson's *Ancient Egyptians* and Lane's *Manners and Customs of the Modern Egyptians*. From the middle of the century guidebooks were incorporated into the canon, and these were no lightweights either. Their publishers prided themselves on including the latest archaeological research, and all of them called on the services of established authorities. Murray's first *Handbook for Egypt* was published in 1847 as a condensed edition of Wilkinson's *Modern Egypt and Thebes*; Baedeker's first *Handbook for Lower Egypt* appeared in 1877 and drew on an array of academic Egyptologists; and in 1886 Thomas Cook commissioned E.A. Wallis Budge, Keeper of Egyptian Antiquities in the British Museum, to prepare a compendious volume of *Notes for Travellers*, which the company presented to all passengers on its Nile steamers. Murray's distinctive handbook was part of the standard outfit of the British tourist, and what Howard Hopley described as 'the inevitable red hand-book' (1869, 281) dominated the market in Egypt until the 1890s; but once Baedeker had combined its separate handbooks on Lower and Upper Egypt in 1898, its German, French and English editions soon led the field (Figure 1). When Francis Maule travelled up the Nile at the turn of the century he was so taken by 'the rosy flush of many and much conned Baedekers' that he used a Baedeker as the cover image for his own, impressionistic account (Maule 1901, 194).

The library also included canonical accounts of travel, like Bartlett's *Nile Boat*, Martineau's *Eastern Life*, Curtis's *Nile Notes* and Edwards's *Thousand Miles up the Nile*. Many of these texts distinguished themselves from the serious monographs by their quasi-apologetic titles or subtitles – 'sketches', 'notes', 'glimpses' – but this did nothing to diminish the knowing style of their prose. Most of them noticed the antiquities, and in some cases offered brief historical summaries and suggested their own routes. But most of these

Figure 1 'Busily reading their Baedekers.' (Charles Dana Gibson)

authors were also interested – usually much more so, in fact – in describing their own experiences and recording their impressions of the people and places they had encountered. These texts formed part of what, following Said, might be called the 'citationary structure' of Orientalism, which Charles Gibson captured perfectly:

> Those who go up the river in a dahabiyeh like to feel that they are in the same boat with the travelers whose books they read from New York to Port Said. This would be a very pleasant feeling, if it did not suggest the responsibility of keeping a record of days that, from all accounts, are sure to be of so much importance. There is a sentimental belief that each day on the river is to be of the greatest importance, just as if thousands of tourists on Cook's steamers were not taking the same journey each year. So overpowering becomes this delusion that even letters home seem to take the form of historic biographies.
>
> (Gibson 1899, 40–1)

These considerations became steadily more important as modernity made inroads into Egypt: being in the same boat as their exemplary predecessors provided ordinary tourists with a precarious guarantee of the authenticity of their own experience.

These writings mapped a double geography. On one side, tourists had to be assured that it was possible to inspect the ancient and the exotic from the comfort and security of the modern, or at least the familiar. There were of course hardy souls who despised what the testy archaeologist Flinders Petrie called 'the superficial class, for whom the tender-foot directions of guide-books are written, and luxuries of hotels are provided as attractions' (1892, 3). But even those of a more adventurous disposition – and none of those who ventured on the Nile in the middle decades of the nineteenth century were exactly timorous – were usually careful to maintain some sort of distance between themselves and local people.[2]

On the other side, tourists also had to be reassured that these incursions of modernity had not destroyed the very object of their trip: that, contrary to the anxieties of belated Orientalism, 'the Orient' was still available for their inspection more or less as they had imagined it.[3] In Egypt this familiar dilemma was double-edged. In the second half of the nineteenth century the progress of archaeological excavation revealed more and more of the ancient past, and brought it within a recognizably modern apparatus of display and classification – an 'exhibitionary complex' – which enlisted local people in subaltern positions as labourers, watch-keepers and guides; but their involvement in these and the many other services installed by modern tourism put at risk the 'authenticity' of the larger experience. And so travel writings were increasingly concerned to put in place a second exhibitionary complex, whose scriptings could provide and police the sights of a (still) exotic Egypt.

EGYPTIAN MODERN: VIEWING PLATFORMS AND
VANTAGE-POINTS

In the first half of the nineteenth century most travel writers rapidly passed over the hotels in Alexandria and Cairo. In the early decades there were none at all for them to recommend, and the choice remained limited until the 1840s. By then there were two main hotels in Alexandria, the Hôtel d'Europe ('Rey's') – which was according to Wilkinson 'the one most frequented by the English' – and the Hôtel d'Orient ('Coulomb's'). In 1839 Emma Roberts had judged them both 'excellent' for 'the accommodation of European travellers', Rey's offering 'every comfort we could desire' (Roberts 1841, 77), but a few years later Bartlett was more grudging: he found them 'rambling and comfortless places', though he did admit that 'all things considered' they were 'surprisingly good for Egypt' (Bartlett 1849, 19). By this time there were also several hotels in Cairo, and in 1843 Wilkinson announced that Hill's Great Eastern Hotel – which had been 'the first hotel for some years' – had been eclipsed by the new Hôtel d'Orient, which was 'said to be very comfortable' and had the advantage 'of more recent improvements' (Wilkinson 1843, 101). He mentioned two others, 'the Giardino or French hotel' – although it had 'no very good rooms' and was 'mostly patronised by French and Italians' – and an Italian hotel which was 'not first-rate' and, he had no doubt, was 'seldom visited by English travellers'. As to the rest, they were 'not worthy of notice' (Ibid., 202–4). Thackeray stayed at the Hôtel d'Orient in 1844, 'as large and comfortable as most of the best inns in the South of France' (Thackeray 1844, 129), as did Martineau and her party. The mainstay of all these hotels was originally the overland trade between Europe and India, but as the Nile voyage became increasingly fashionable so hotel-owners started to look to the tourist trade. By 1850 the Great Eastern had become Shepheard's Hotel and moved into a former royal palace overlooking the Ezebekieh, and several other hotels had started operations. In both Alexandria and Cairo all of the recommended hotels were in the 'Frank' quarter, and travel writers took it for granted that they would be under Western management and hence evaluated by Western standards.

Travel writers offered particularly detailed advice on the *dahabeeah* voyage, which was constructed as the defining experience of tourism in Egypt during the middle decades of the nineteenth century. Some of their prescriptions were themselves codified in other texts. Great importance was attached to securing the services of a reliable dragoman, a local interpreter-guide. In Cairo, the base camp for the ascent of the Nile, dragomans would present prospective employers with packets 'of dirty, thumbed, faded and torn letters of recommendation' (Beaufort 1861, 41), but tourists were warned to treat them with the greatest circumspection. They were 'often so carefully worded that very contrary meanings may be deduced' (A daughter of Japhet 1858, 51), and in some cases testimonials were simply 'passed from one [dragoman] to another' (Eames 1855, 26).[4] The hire of the boat was also formalized in a written contract. 'I give here our contract verbatim,' Prime told his readers (1857, 121–3), while

Warren, who did exactly the same, advised that 'the contract should be always made in duplicate and the signatures attested at the consulate' (Warren 1883, 17–20).

Other advice was issued in elaborate lists: books to be read in advance and books to be packed for the journey, supplies to bring from home and supplies to obtain in Cairo, and last – but never least – a veritable chest of drugs and medicines. The *dahabeeah* would usually be furnished, but most travellers added their own touches. They supplied their own blankets and linen, and often bought additional chairs and carpets in Cairo. Nor was this all: 'The ladies of the party', Hoskins advised, 'may make coverings for divans and twenty-four cushions, and curtains for windows and doors' (1863, 91). He thought that sixty yards of chintz would be adequate for the purpose. Edwards described her dining saloon in detail, with its panelled walls, its rich carpet and its curtains:

> Add a couple of mirrors in gilt frames; a vase of flowers on the table . . . ; plenty of books; the gentlemen's guns and sticks in one corner; and the hats of all the party hanging in the spaces between the windows; and it will be easy to realise the homely, habitable look of our sitting-room.
>
> (Edwards 1877, 40)

The object of all this was to fashion a floating home away from home. Thus Bartlett described his *dahabeeah* as 'a world in itself': 'I was monarch of all I surveyed,' he declared proudly, 'and amused myself with arranging everything in the nicest order' (1849, 109). In Warren's eyes, 'everything appeared truly oriental, except our saloon and table. They looked so cheerful and Parisian that we felt at home at once' (1883, 25). And on the *dahabeeah* hired by Prime and his wife 'the furniture was oriental, of course: but two American rocking-chairs made things look somewhat natural within the cabin' (Prime 1857, 124).

These stage directions transformed the *dahabeeah* into a secure viewing platform from which travellers could inspect and on occasion issue out into 'the Orient' (Figure 2). The contrast between the two was indelibly inscribed in travel accounts of the period. Bayard Taylor, who travelled up the Nile into Nubia and Ethiopia in 1851–2, described his boat as 'a floating speck of civilisation in a country of barbarians' (Taylor 1854, 51). 'Into the heart of a barbarous continent and a barbarous land,' he wrote in his journal, 'we carry with us every desirable comfort and luxury' (Ibid., 96). Much later, but in much the same spirit, Elizabeth Butler recalled a trip made in the winter of 1885–6:

> Travelling thus on the Nile you see the life of the people on the banks, you look into their villages, yet a few yards of water afford you complete immunity from that nearer contact which travel by road necessitates; and in the East, as you know, this is just as well.
>
> (Butler 1909, 55)

Figure 2 Dahabeeah at Philae, Upper Egypt in 1856. (Francis Frith)

That said, many travellers insisted that the *dahabeeah* allowed for an intimacy of sorts and, indeed, that it was precisely this closeness that gave the experience its authenticity. This was increasingly valued as the nineteenth century wore on and the incursions of a European modernity became ever more apparent. As Edwards explained,

> In Europe, and indeed in most parts of the East, one sees too little of the people to be able to form an opinion about them; but it is not so on the Nile. Cut off from hotels, from railways, from Europeanised cities, you are brought into continual intercourse with natives. The sick who come to you for medicines, the country gentlemen and government officials who visit you on board your boat and entertain you on shore, your guides, your donkey-boys, the very dealers who live by cheating you, furnish endless studies of character, and teach you more of Egyptian life than all the books of Nile-travel that ever were written.
>
> Then your crew, part Arab, part Nubian, are a little world in themselves. . . .
>
> (Edwards 1877, 166–7)

Any intimacy with those other 'little worlds' was always conditional, however, partly because of difficulties of communication – in most cases each had at best a rudimentary command of the other's language[5] – and partly because the tourist could always terminate any engagement and substitute the languid gaze, peremptory instruction or even physical punishment for 'continual intercourse'. In any event, all interaction was hedged around with restrictions laid out in those 'books of Nile travel' and, as we will see, Edwards was by no means immune to them. Travel writers provided detailed recommendations on dealing with the dragoman, *reis* (captain) and crew on board, and with the local officials and villagers encountered en route.

Largely ignorant of language and even custom, most tourists were more or less completely dependent on their dragoman; but this reliance produced an unease, even a distrust that could spill over into outright contempt. When Curzon notes that the traveller 'sees through his dragoman's eyes, hears through his ears' (1849, 66) and Curtis objects that dragomans 'travel constantly the same route, yet have no eyes to see nor ears to hear' (1858, 112), these are more than different impressions of different experiences. They reveal a basic ambivalence in the relationship, and one which travel writers tried to adjudicate in advance. The dragoman was assumed to act purely in his own interest, and his local knowledge was typically marginalized by attributing it to guile, malice or simple ignorance against which the tourist had to prevail. 'Things contrary to custom need a certain amount of insistence,' Edwards urged, 'and are sure to be met by opposition.'

> No dragoman, for example, could be made to understand the importance of historical sequence [to sight-seeing on the Nile]. . . . To him, Khufu, Rameses and the Ptolemies are one. As for the monuments, they are all ancient Egyptian, and one is just as odd and unintelligible as another. He cannot quite understand why travellers come so far and spend so much money to look at them; but he sets it down to a habit of harmless curiosity – by which he profits.
>
> (Edwards 1877, 70)

It was the same when it came to bargaining, which was full of pitfalls for the unwary – and not only in the frenetic bazaars of Cairo. Tourists were urged to be particularly careful of the sellers of antiquities since they were often also the manufacturers of 'antiquities'. When Bartlett visited Luxor in 1845 he was offered 'scraps of papyrus and mummy cases, coins, scarabei, &c.' which he thought 'most suspiciously modern in appearance' (Bartlett 1849, 148–9). Ten years later Prime warned tourists about Luxor's principal dealer in antiquities, 'Ibrahim the Copt':

> While he is confidentially informing you [about the manufacture of counter-feit antiquities] . . . how this is modern and that is not . . . beware, lest you become too trusting and he sells you in selling a ring or a vase or a seal. He is a wily fellow and sharp, and he knows well how to manage a Howajji [foreign traveller].
>
> (Prime 1857, 216–17)

The business was still thriving in the 1860s. Hoskins found a lively 'traffic in antiquities', much of it conducted by the local Consuls at Luxor, who he said 'sell now at immense prices what cost them little or nothing'. 'When such prices are given,' he continued, 'it is not surprising that a manufacture of them has sprung up' (Hoskins 1863, 218–19).

However acceptable such an appeal to the logic of the market might have been at home, tourists were markedly irritated by its extension into Egypt. They complained of rising prices but, even more, of the ways in which local people had absorbed the lessons of the modern market-place. Tourists continued to be annoyed by traditional demands for 'baksheesh', but it was the mastery of values, prices and exchange rates that really got their goat. As Smith put it, local people 'have learnt to become more and more extortionate and rapacious every year' (Smith 1868b, 3). 'Extortion', so it seems, was a one-way street: it was perfectly acceptable for a tourist to find 'a bargain' in the bazaars, but unthinkable for local people to be allowed a similar advantage. By the early 1870s these economic skills had been sharpened throughout Upper Egypt. Edwards was astonished – even alarmed – by the persistent attempts to sell her something (anything) whenever she ventured into a village in Upper Egypt, but even more by the knowledge of the sellers:

> [T]he women not only know how to bargain, but how to assess the relative value of every coin that passes current on the Nile. Rupees, roubles, reyals, dollars and shillings are as intelligible to them as paras or piastres. Sovereigns are not too heavy nor napoleons too light for them. The times are changed since Belzoni's Nubian, after staring contemptuously at the first piece of money he had ever seen, asked 'Who would give anything for that small piece of metal?'
>
> (Edwards 1877, 239)[6]

In these and many other ways it was widely acknowledged that local people had become skilled participants in the expanding networks of travel and tourism, and that they had indeed learned 'how to manage a Howajji'. 'Taking strangers up the Nile seems to be the great business of Egypt,' Warner warned his readers, and for those strangers – 'innocents abroad', as Mark Twain famously called them (1869) – 'all the intricacies and tricks of it are slowly learned' (Warner 1876, 52–3). 'Trickery' again, but it implied something other than pure deceit. When Arnold described tourists and local people making moves on 'the chess-board of dahabeeah travelling', he was configuring tourism as an elaborate game, a battle of wits, which was supposed to be part of the fun (Arnold 1882, 157). 'The traveller who thinks the Egyptians are not nimble-witted and clever,' Warner cautioned, 'is likely to pay for his knowledge to the contrary' (Warner 1876, 92). A game it may have been, but there was also an underlying seriousness of purpose, and travel writers constantly emphasized the difference between – and the importance of regularly *reminding* people of the difference between – the kings and queens and the pawns. If chess originated in the East, to continue the metaphor, the rules of the modern game were to be those of the West. Taylor captured this scripting in an arresting image:

> We are on board our own chartered vessel, which must go where we list, the captain and sailors being strictly bound to us. We sail under national colors, make our own laws for the time being, are ourselves the only censors over our speech and conduct, and shall have no communication with the authorities on shore, unless our subjects rebel. Of this we have no fear, for we commenced by maintaining strict discipline, and as we make no unreasonable demands, are always cheerfully obeyed. Indeed, the most complete harmony exists between the rulers and the ruled, and though our government is the purest form of despotism, we flatter ourselves that it is better managed than that of the Model Republic.
>
> (Taylor 1854, 87)

The reversal of the usual gibe about 'Oriental despotism' served to confirm the power of the foreign traveller as what Warren, like so many others, called 'a perfect monarch on his dahabeah, master of her movements' (1883, 6).[7]

The rule of precedence was thus perfectly clear to most tourists – 'a whole squadron of [native] boats was waiting at the locks [at Afteh]', Taylor wrote, 'but with Frankish impudence we pushed through them and took our place in the front rank' (1854, 26) – but they frequently disputed its national inflections among themselves. 'There is an imperiousness in the English mind', Frances Cobbe explained:

> Assuredly we owe it to the way in which travelling is facilitated in every corner of the globe where English people do congregate. One after another we pour on, staring at every delay, insisting on more and more rapid conveyance, fretting, fuming, making ourselves objects of astonishment to the calm Oriental and of ridicule to our fellow Europeans; but still eventually always conquering and leaving rough places smooth and crooked things straight behind us.
>
> (Cobbe 1864, 43–4)

This sort of attitude did nothing to contain national rivalries, of course. Disputes between the French and the English over their respective roles in Egypt were a commonplace of international diplomacy right through the nineteenth century, and their tourists dutifully lined up on one side or the other. However, as the century wore on and more American and German tourists appeared in Egypt so they too were drawn into the circles of chauvinistic bickering. Here, for example, is one young American who, even while he was on his way to Alexandria, was already keenly aware of the emerging stereotypes:

> This was my first intimate acquaintance with Germans as tourists, and I never want to hear again about the 'loud, self-asserting boisterousness' of

Americans abroad. Why, these people monopolized the ship; they monopolized the steamer-chairs; they monopolized the conversation; and if the food
hadn't been so bad they would have monopolized that.

(Reeve 1891, 20)

The culture of caricature and complaint was widespread, but these were all intra-
mural disputes, leaving no doubt – at least in the minds of the tourists – that they
moved in and presided over separate worlds.[8]

The tourist's sovereignty was also conveyed through visible markers. Each
party of tourists renamed their boat for the duration of the voyage and equipped
themselves with a personal pennant; these were originally registered with consular
officers in Cairo, but from 1849 most tourists simply copied them into a large
ledger kept for the purpose at Shepheard's Hotel. They also sailed under their
own – often extravagantly large – national flag.[9] These various signals were part of
an elaborate scripting of social conventions among those travellers making up
what Curtis called 'the society of the River' (1858, 71). 'Passing each other by
day,' Edwards wrote, 'we dip ensigns, fire salutes and punctiliously observe the
laws of maritime etiquette' (Edwards 1877, 39). Such courtesies were usually
only extended to other 'Franks', however, and never to the hundreds of ordinary
native boats plying the river. There was a strict micro-geography on board the
dahabeeah too. As Edwards also explained, the upper deck 'is the exclusive territory of the passengers' while 'the lower deck is the territory of the crew' (Ibid.,
91).[10] Another writer elaborated:

> The crew is not allowed to walk across the upper deck but must always walk
> outside the boat on the ledge which is placed purposedly all round it. It is
> very necessary to adhere strictly to this rule, for otherwise the steersman,
> captain, cook-boy and one or two sailors are perpetually passing by to helm
> or bow on one excuse or another, which is extremely unpleasant and by no
> means safe as regards cleanliness.

(Carey 1863, 86)

There were also explicit directions about appropriate costume. Martineau
recommended brown holland as the best material for ladies' dresses: 'nothing
looks better if set off with a little trimming of ribbon.' Caps and frills were perfectly useless, she said, but round straw-hats with a broad brim, together with
goggles of black woven wire, were indispensable protection against the sun. Her
one indulgence was to allow the tarboosh as 'becoming head-gear', but only
'when within the tent or cabin' (Martineau 1848, 43). It was vital to make as few
concessions as possible: 'It is better [for an Englishwoman] to appear as she is,'
she declared firmly, 'at any cost, than to attempt any degree of imposture' (Ibid.,
521). In the middle decades of the nineteenth century men were more likely than
women to follow Taylor's example and, as he put it, 'un-Frank' themselves, especially if they were travelling on their own or with a male companion. Taylor
performed a teasing change of costume for his readers, in which he first adopted

the *tarboosh* and white turban; then the 'flowing trowsers and embroidered jacket', which he found 'easy and convenient in every respect' (Taylor 1854, 92–3); finally, with the addition of a light silk shirt he 'assumed the complete Egyptian costume', which 'in its grace, convenience, and adaptation to the climate and habits of the East' was 'immeasurably superior to the Frank costume. It allows complete freedom of the limbs, while the most sensitive parts of the body are thoroughly protected from changes of temperature' (Ibid., 169).[11] By the 1870s the dress code for most men had become as uncompromising as that for women. Two tweed suits were the best outfit, Murray's *Handbook* firmly advised, together with a felt hat with white muslin wrapped around it or even a pith helmet. But the 'red tarboosh with which travellers so often delight to adorn themselves' was frowned upon: 'no respectable European resident in the country would think of appearing in it in public' (*Handbook* 1873, 6).

In all these ways, then, lines were drawn and attempts made to police the inevitable interactions across them. In this respect it is salutary to read Lady Gordon's opinion of Martineau's *Eastern Life*. Gordon lived at Luxor for many years, and her letters home reveal a profound attachment to the local people and an increasingly critical attitude towards tourists who sailed uncomprehendingly by scenes of extraordinary poverty and privation. Her verdict is thus scarcely surprising. 'The descriptions are excellent,' she concluded, but the book suffered from what she saw as 'the usual defect – the people are not real people, only part of the scenery to her, as to most Europeans.' Gordon had no doubt about the reason: like other foreign birds of passage, Martineau felt 'that the difference of manners is a sort of impassable gulf' (Gordon 1969, 120).[12]

Towards the end of the nineteenth century the scope of the Egyptian modern was enlarged and that 'impassable gulf' widened: the imaginative distance increased between the modern viewing platforms of the tourist and the lifeworlds of local people. This was, in part, a matter of physical provision – the construction of grand hotels and fleets of tourist steamers – but it also depended on the ways in which these new spaces were scripted. Alexandria had long been regarded as a part-European city, 'a piebald town' Warburton had called it (1844, 17), but during the late 1860s the francophile Khedive Ishmail embarked on a grand design to transform Cairo into a 'Paris-on-the-Nile'. He contracted with French architects, planners and landscape gardeners to establish a new quarter of wide boulevards, public parks, fine villas and luxury hotels. Foreign opinion was sharply divided. Traditionalists sneered at the invasions of a European modernity – in a vivid reversal of the usual feminization of the Orient, Georg Ebers declaimed that 'Civilisation, that false and painted daughter of the culture of the West, [had] forced her way into Egypt' (Ebers 1879, iv) – while others were prepared to grant the value of the changes. 'There is no country in Europe where one could live in such perfect luxury as during six months in Egypt,' wrote Leland, 'were there only a really first-class city of comfort there.' And by 1873 he could see the signs of its triumphant emergence: 'What a place for a Paris!' (1875, 84).

By the turn of the century Cairo was 'emphatically a society place,' announced Reynolds-Ball, and 'as an aristocratic winter-resort it ranks with Cannes or Monte

Carlo' (1907, 44–55). He distinguished three categories of hotel in the city and offered his readers a series of social discriminations that were also social expectations. Within the fashionable tier of grand hotels, he judged that the Savoy was 'more peculiarly exclusive and aristocratic, while the Continental and Shepheard's are smarter and the note of modernity more insistent' (Figure 3). He described Shepheard's clientele as 'distinctly cosmopolitan' – which turned out to mean that it had become 'the American hotel' – while the Savoy 'is more exclusively English' (Ibid., 124–9). The same social tone prevailed on the first-class steamers, which had come to resemble late Victorian and Edwardian house parties. Here is Sladen, the snob's snob, explaining that 'Cook's boats are like the best hotels':

> There is the same boy with the ostrich-feather broom waiting to dust your legs and feet; the same procession of Arab porters in gowns waiting to seize your luggage. Cook's European stewards, mostly Italians, are very superior to the European waiters you get in the hotels; they attend to you instead of to

Figure 3 Shepheard's Hotel, Cairo.

themselves. I wonder that there are not people who spend their lives on Cook's Nile steamers, whenever they are running, just as they spend their lives at golf-clubs which have bedrooms. Here you have the *dolce far niente* materialised. The dining-room is on deck, and full of windows; you can see the scenery while you are eating. The fore-end of the saloon is one gigantic window. You can also see the scenery as you lie in bed through the big window of your cabin. . . . Instead of bunks you have good high beds, high enough from the ground to take two or three ladies' dress-baskets under-neath – a consideration in a place where people change their clothes as often as they do in the smart society of Cook's boats.

(Sladen 1910, 388–9)

In such circumstances there were absolutely no concessions to any tourists tempt-ed to 'go native': 'The particular young man sometimes breaks out into silk suits and wonderful socks,' enthused Sladen, 'or at any rate rare and irreproachable flannels, just as the young girl who has come to conquer Cairo society rings the gamut of summer extravagances' (Sladen 1908, 425). Shore excursions required different attire, and Lorimer acerbically noted that on such occasions 'few Europeans know how to dress in Egypt so as not to offend the eye and disfigure the landscape' (Lorimer 1909, 130). But it was essential to avoid caricature: 'Helmets and "puggarees", while picturesque, are by no means essential to com-fort, and if worn in winter they are the shameless insignia of the guileless tourist' (Marden 1912, 14–15).

As that last remark indicates, tourists were still advised to be on their guard. The bazaars in Cairo were full of 'so-called Oriental articles' that were manu-factured in Manchester, Baedeker warned (1898, 28), while it was business as usual at Luxor: 'Half the population is engaged in traffic with antiquities, and the practice of fabricating scarabei and other articles frequently found in tombs is by no means unknown to the other half' (Ibid., 226). It was impossible to protect every tourist from deceptions like these, but many tourists sought security through the simple expedient of joining an organized party or even a conducted tour. There were several companies operating in Egypt by the end of the century, but it was the British firm of Thomas Cook & Son that soon dominated the market (see Bredon 1991; Withey 1997). 'The very name of Cook', Ward announced, 'becomes in Egypt a magic talisman, securing all who trust in it immunity from fraud and protection from rudeness, incivility [and] petty annoy-ances of any kind' (Ward 1900, 173). Once on board the tourist steamer, the terms of engagement were perfectly clear: local people would be called on to perform as stage-hands – suitably trained servants who would attend to the tour-ist's every need – or as actors in a spectacle devised purely for the entertainment of the tourist.

'Messrs Cook take an ignorant Arab or an ebon-tinted Nubian from his native village,' Ward wrote with evident approval, 'put him through some mysterious training, known only to themselves, and in a short time he is fit for use, is labelled "Cook" in large letters, and lo! he at once becomes a patient, efficient and trustworthy servant of all bearers of their tickets' (Ibid.). It seemed to work. 'The servants on Cook's boats spoil you for any other servants,' Sladen marvelled. 'They hang about you like shadows in soft white robes, wondering what you could want next.'

> Tea is not laid on Cook's boats; these white-robed spirits hover round you with tea-pots and milk-jugs and sugar-basins, and a dozen different kinds of Huntley & Palmer's biscuits. In the same way at meals they notice what tit-bits you like, and observe your idiosyncrasies in the arrangement of your toilet-requisites round washing-stand and mirror. Is there anyone who is not particular about the place occupied by his toothbrush when at rest?
>
> (Sladen 1910, 391)

The 'white robes' were regarded as part of the requisite effect. In much the same way, once he had got over the pleasures of his toothbrush, Sladen was thoroughly delighted to find that the chief dragoman on board was 'Mohammed, the *doyen* of Cook's', not only because he conjured up what Sladen was pleased to call 'the good old days' – though he was now, of course, answerable to the company for his actions – but also because he was 'even more endeared to the tourist by the picturesqueness of himself' (Sladen 1908, 427). 'Picturesqueness' – what Sladen elsewhere called the 'humours' of native life – was an absolute requirement; it was essential for the actors to look the part. Lorimer echoed the sentiments of many of her companions when she declared she would be perfectly satisfied 'if there was nothing else to watch all day but the antics of our Soudanese crew':

> There is the ostrich-feather broom boy who watches for a speck of dust to brush away, and the brass boys who lift up rugs and mats to find some hidden treasure in the way of knobs to polish, indeed there is a boy with a grinning smile and flashing teeth for every mortal occupation you can imagine. I often wonder if there is a special crew kept to do nothing but say their prayers, for there is always a group of black skinned Soudanese in white drawers on their knees in the bows of the boat. Perhaps Thos. Cook and Son recognise how valuable they are for 'off days' on the Nile, for tourists to kodak.
>
> (Lorimer 1909, 169)

Several writers archly referred to the ways in which assumptions like these produced a new space of sovereignty. Steevens reported that the 'nominal governor' of Egypt was the Khedive but that its 'real governor' was Thomas Cook & Son (Steevens 1898, 68), while still more extravagantly Sladen declared from the deck of his steamer that 'Cook is the uncrowned King of Egypt, and this is the navy with which he won his battle of the Nile' (Sladen 1910, 388). But these supposedly comic claims had more than metaphorical force. The scripting of these new spaces not only required local people to approach on the tourist's own terms: when tourists ventured into the world of ordinary Egyptians this was understood to be a short-lived adventure, a lark, and that earlier, extraordinary sense of detachment – the ability of the tourist to move in and out of 'the Orient' at will – was reasserted in even more powerful forms.

In Cairo F.C. Penfield captured the juxtaposition precisely when he observed how

> The traveller of impressionable nature yields to the fascination of Cairo's quaint Eastern life, as perfect as if met far beyond the Orient's threshold, and doubly satisfying because found within a half-hour of the creature comforts of hotels conspicuously modern. To walk the streets of an Oriental capital wherein history has been made, between meals, as it were, and delve by day in museums and mosques perpetuating a mysterious past, and dine *de rigeur* in the evening, with the best music of Europe at hand, explains a charm that Cairo has for mortals liking to witness Eastern life provided they are not compelled to become part of it.
>
> (Penfield 1903, 2)

Butler said much the same – 'You may hide in the bazaars, but you cannot live there' (1909, 36) – and recalled her excitement at

> diving into the old city, and in a ten minutes' donkey-ride [finding] oneself in the Middle Ages; in the real, breathing, moving, sounding life of the Arabian Nights. . . . Then when inclined to come back to our own time and comforts, which I am far from despising, ten minutes' return ride and the glimpse into the old life of the East became as a vision.
>
> (Butler 1909, 39)

It was exactly the same on the tourist steamer. 'A vision of half-barbarous life passes before you all day,' Steevens observed happily, 'and you survey it all in the intervals of French cooking': 'Rural Egypt at Kodak range – and you sitting in a long chair to look at it' (1898, 215).

EXCAVATION AND EXHIBITION

'Alexandria', reported William Curtis in 1905, 'is a city of sites instead of sights, which is a clever epigram and almost true, because you can only see the places where great historical structures once stood' (Curtis 1905, 20). For precisely that reason Alexandria, like Lower Egypt as a whole, was rapidly passed over by most tourists, whose attention was immediately drawn to the monuments of Cairo and Upper Egypt. Many of them were obvious, but the production of other sights was dictated by the uneven progress of archaeological research. By the 1840s most of the major sites had been identified. When John Gardner Wilkinson's *Manners and Customs of the Ancient Egyptians* appeared in 1837, the third volume included an appendix based on notes he had drawn up for friends 'who required a brief statement of the principal objects worthy of a visit on the Nile,

without having to seek their order and position in the numerous pages of volu-
minous books of travel'. To that end Wilkinson simply listed the main sites and
gave terse directions to them (1837, 399–404), but in 1843 he decided that a
detailed handbook was 'indispensably necessary'. His *Modern Egypt and Thebes*
combined an account of the antiquities with a compilation of 'the information
required for travellers' (Wilkinson 1843). In 1847 Wilkinson's publisher, John
Murray, revised and reorganized the text to conform with the model he had
established for his own series of European handbooks. The sites were organized
into 'routes' and the new version was published as Murray's first *Handbook for
Travellers in Egypt*.

If the main sites were known to the travelling public, however, their excavation
was still in its infancy. In 1848 Martineau wished for

> a great winnowing fan, such as would . . . blow away the sand which buries
> the monuments of Egypt. What a scene would be laid open then! Who can
> say what armies of sphinxes, what sentinels of colossi, might start up on the
> banks of the river, or come forth from the hill-sides of the interior, when the
> cloud of sand had been wafted away!
>
> (Martineau 1848, 45)

Many other travellers, encountering ruins half-buried in sand, choked with deb-
ris, or packed with the mud huts and pigeon-towers of a native village, must have
had the same thought. It did not sensibly diminish as excavation proceeded. 'We
wondered how much still lay buried [at Memphis],' wrote the Tirards in 1891,
'and how long it would be before the spade of the excavator would unearth the
hidden treasure' (Tirard and Tirard 1891, 8).

Martineau had been quick to rein in her fantasy. 'It is better as it is,' she had
concluded, 'for the time has not come for the full discovery of the treasures of
Egypt' (Martineau 1848, 46).[13] But even as she wrote the world was turning.
In 1850–1 the young French archaeologist Auguste Mariette discovered the
Serapeum at Memphis, and on his return to Egypt in 1857 he directed a veritable
frenzy of excavation at a number of other sites, including Gizeh, Abydos and
Thebes. His discoveries confirmed the primary position of these sites within
the emerging hierarchy of 'sights', and Curtis had already foreseen the
consequences:

> [E]xcavation implies cicerones and swarms of romantic travellers in the way
> of each other's romance. . . .
>
> Excavation implies arrangement, and the sense of time's work upon a tem-
> ple or a statue, or even a human face, is lost or sadly blunted, when all the
> chips are swept away, and his dusty, rubbishy work-shop is smoothed into a
> saloon of sentiment. . . . When it is not enough that science and romance
> carry away specimens of famous places to their museums, but Mammon
> undertakes the making of the famous place itself into a choice cabinet, they
> may be esteemed happy who flourished prior to that period.

And it is pleasant to see remains so surpassingly remarkable, without having them shown by a seedy-coated, bad-hatted, fellow-creature, at five francs a day.

(Curtis 1858, 326–7)

This was less a restatement of the romantic attachment to ruins than a timely recognition of its precariousness.[14] As his appeal to 'Mammon' shows, Curtis realized that the production of sights was also the production of commodities – the exhibition of antiquities in a 'choice cabinet' – and he was bleakly realistic about where the process would end:

[T]hese things will come. Egypt must soon be the favourite ground of the modern Nimrod, travel – who so tirelessly haunts antiquity. After Egypt, other lands are young, and scant, and tame, save the Parthenon and Pestum. Every thing invites the world hither.

It will come, and Thebes will be cleaned up and fenced in. Steamers will leave for the cataract, where donkeys will be in readiness to convey parties to Philae, at seven A.M. precisely, touching Esne and Edfoo. Upon the Libyan suburb will arise the *Hôtel royal au Rameses le grand* for the selectest fashion. There will be the *Hôtel de Memnon* for the romantic, the *Hôtel aux Tombeaux* for the reverend clergy, and the *Pension Re-ni-no-fre* upon the water-side for the invalides and sentimental – only these names will then be English; for France is a star eclipsed in the East.

(Curtis 1858, 327–8)

This was uncannily accurate. In the course of the next fifty years or so the remains of ancient Egypt were staged and scripted as sideshows in an extended exhibition. This had three central elements: the formation of a museum, the production of the 'sights', and the compilation of guidebooks. By the middle of the nineteenth century museums in London, Berlin and Paris had impressive Egyptian collections which helped to shape the expectations of many tourists. In Cairo Mohammed Ali and his adopted son Ibrahim maintained private collections of antiquities, but they were so unsystematic that Wilkinson considered 'the formation of a museum in Egypt [to be] purely Utopian' (1843, 265). In 1854, however, the Museum of Egyptian Antiquities was established in the old post office at Boulaq. Conditions were far from ideal – the building was too small and always damp: on some mornings shrouds of fog drifted off the river and trailed through its corridors – but it was not until 1889, when Egypt was under British occupation, that the collection was moved to a much larger though still makeshift space in the Gizeh Palace. Soon after, work started on the construction of a modern museum in the new quarter of the city (Wallis Budge 1906, 426–8). It opened in 1902 to enthusiastic applause from travel writers: 'Egyptologists [claim] that in point of classification of the objects collected here the Museum may serve as a model to most of the great museums of Europe' (Reynolds-Ball 1907, 77).

The exhibition extended far beyond the walls of the museum. By the closing decades of the nineteenth century the major sights had all been 'cleaned up and fenced in', just as Curtis had predicted, and turned into 'standing show-places for tourists' (Ferguson 1873, 70). 'From Memphis to the gates of the Sudan,' wrote Sladen, 'Egypt is an open-air museum where temples and tombs are arranged like shop windows for public inspection' (Sladen 1911, 332). The government required tourists to buy a ticket which covered admission to the principal antiquities in Upper Egypt. Travel writers now referred to 'the round of regulation sights', and at each site there was a standard route. Even the scholarly Wilkinson had offered 'instructions for the quickest mode of seeing the objects most worthy of notice' at Thebes 'and the order in which they may be visited'. Perhaps the prospect secretly scandalized him; it certainly appalled later and far less erudite critics. But Wilkinson was a pragmatist. 'The traveller who merely wishes to *say he has seen* Thebes', he conceded, and many plainly did, 'may get through it all in three days' (1843, 135–6). Subsequent writers followed his example. The steamers allowed three days at Thebes, and dragomans and donkey-boys soon learned the drill: one visitor reported that they 'know these tombs only by the numbers Sir Gardner Wilkinson has assigned them' (Arnold 1882, 279). Routes were forcefully prescribed for other sites too. 'The route is indicated by a line on the map,' Baedeker instructed visitors to Philae, 'and any deviation suggested by the dragoman should be rejected' (Baedeker 1898, 339). The staging was completed by the installation of electric lighting. In the past tourists had groped their way through the temples and tombs with only spluttering candles or magnesium flares to guide them: 'Now', one of them marvelled, 'it is like entering a modern ballroom' (Curtis 1905, 176). The dispersed exhibition was crowned by Baedeker, who awarded the most important sights one or two 'asterisks' (stars) as 'marks of commendation'.

The passage of tourists through the exhibition was scripted by guidebooks. The function of Murray or Baedeker was to bring the exhibits within an intellectual landscape that established their significance, and to bring the tourist-reader within a text the transparency of which revealed 'an objective geography' (cf. Grewal 1996). This double movement worked through a logic of anticipation and confirmation. Baedeker's characteristic manoeuvre was to travel ahead of and alongside the tourist, thus:

> On the left bank, as we draw near Thebes, rise high limestone hills, presenting precipitous sides to the river. . . . The right bank is flatter, and the Arabian hills retreat farther into the distance. Before reaching the point where the W. chain projects a long curved mass of rock towards the river, we see to the left first the great obelisk, and the pylons of the temple of Karnak, half-concealed by palm trees. When we clear the abrupt profile of the W. cliffs . . ., we may catch a distant view of Luxor towards the S.E. None of the buildings on the W. bank are visible until the steamer has ascended as high as Karnak; then first the colossi of Memnon and afterwards the Ramasseum and the Temple of Der el-bahri come into view. The telegraph-posts and wires, which here

obtrude themselves upon the view, seem strangely out of place beside the majestic relics of Egypt's golden period. As we gradually approach Luxor, we distinguish the flags flying above the white houses on the bank and from the consular dwellings. . . . In a few minutes more the steamer halts, close to the colonnades of the mighty temple of Amenophis III.

(Baedeker 1898, 225)

The effect of this cinematographic gaze was extraordinarily powerful. 'Is there not a charm', one tourist asked, 'in knowing that some city, some temple, some natural feature you have tried to realize in your mind is about to appear in very truth just round that bend of road or river?' (Butler 1909, 56). It was more than charm. The correspondence between the description and its appearance 'in very truth', each the projection of the other, was achieved by having the text organize and track the view: the tourist anticipated what was about to be seen, because Baedeker already knew it, and the scene then unfolded according to the compositional logic of his description, which directed attention first this way and then that. The matter-of-factness of the prose, clearing away the clutter of the telegraph poles and gradually bringing objects into focus until the tourist-reader was brought right up to the colonnades of the temple, worked to establish the very facticity – the 'objectivity' – of the sight itself.

Confirmation – 'the will to verify' mobilized by the guidebook (Behdad 1994, 44) – was achieved by both reading and recording the scene. In 1873 Murray still reprinted an annotated version of Wilkinson's original list of 'certain points requiring examination' which had been directed towards 'those who are induced to make researches'. Even in Wilkinson's day many of these points must have been much too sharp for most travellers: 'Excavate, if possible, the site of the temple of Heliopolis' (Wilkinson 1843, 417–19). But the other suggestions were also exacting: 'Copy all the fragments of inscriptions on the numerous blocks of granite [at Sân]'; 'Copy the whole of the inscriptions in some tombs of the Old Empire'; 'Make a plan of the old Roman fortress and of the Coptic Church [in Old Cairo]'; 'Copy the great hieroglyphic inscription on the right of the entrance to the grotto [at Asyoot]' (*Handbook* 1873, 43–4). How many tourists responded to such directives is conjectural – some certainly did (Smith 1868a, 208; Edwards 1877, 336) – but the emphasis on copying was part of a wider culture of inscription. Drawing sketches, painting watercolours, taking photographs, and above all recording information and impressions in diaries and notebooks, served not only to authenticate the presence of the tourist at the scene – and thereby confer representational authority – but also to confirm that from such a viewpoint the jumble of ruins in the landscape could be brought to a visual and textual order. Bartlett described 'sitting in the little cabin, reducing to order the sights of the day' (1849, 19), for example, while Eames noted the obligation 'to see by day, and at night, record what I have seen'. 'When the day's excursion is over,' she wrote, it is important 'to think over each point, to be sure that you have everything arranged in your mind, in its proper order' (Eames 1855, 127).[15] Murray and Baedeker played an important part in bringing the sights to book, as it were, and experienced travellers like Harry Dunning recommended that tourists re-read their guidebooks after an excursion 'to fix what you have seen firmly in the memory' (1905, 36, 103). Many tourists were tempted to do more than merely fix the scene in their minds. Georges Montbard lampooned the 'febrile rustling of pages' as Cook's tourists busily turned over 'the leaves of their guide-books in search of the descriptive note which, slightly mutilated and enlarged by their personal impressions, was inscribed, after due meditation, and with a thoughtful air, in their note-books' (1894, 2–3). Allowing for Montbard's evident condescension, this is perhaps the logical terminus of that dialectic of anticipation and confirmation which scripted modern sightseeing.

If all this makes sightseeing seem like work, and routine work at that, its burden did not lessen as the nineteenth century wore on. As tourists attempted to see more of Egypt in less time, so what Appleton called 'the bondage of Murray' (1876, 63) came to be endured rather than enjoyed. The guidebook's panoramic gaze surveyed the scene so exhaustively that it set the tourist a Sisyphean task: 'It is dreadful hereabouts to read Murray, for he tells long stories of tombs and temples everywhere which we cannot contrive to see. It is the punishment of Tantalus to read of pictures we shall never see' (Ibid., 272). The sentiment was widely held, but it was not just regret at leaving 'more items in the guidebook to be marked as seen' (Dunning 1905, 103). It was the compulsion and regimentation of it all that wearied many tourists, and Warner was surely not alone in his dream of visiting 'an interesting city where there are no sights'. For, as he explained,

> That city could be enjoyed; and conscience . . . would not come to insert her thumb amongst the rosy fingers of the dawn . . . [to suggest] 'Today you must go the Pyramids', or 'You must take your pleasure in a drive in the Shoobra road' or 'You must explore dirty Old Cairo'.
>
> (Warner 1876, 72)

The impersonal prose and imperious commands of the guidebooks could not compensate for the breakneck speed of what Lee Bacon satirized as the 'Rameses Limited Express' – on the contrary, they contributed to it – and for this reason many tourists elected to complement Murray and Baedeker with less prosaic travel writings (Bacon 1902, 155–9). As Edwards argued, those who limited themselves to 'mere guide-book knowledge' were content 'to read the Argument and miss the Poem'.

> In the desolation of Memphis, in the shattered splendour of Thebes, [the tourist] sees only the ordinary pathos of ordinary ruins. As for Abou Simbel, the most stupendous historical record ever transmitted from the past to the present, it tells him but a half-intelligible story. Holding to the merest thread of explanation, he wanders from hall to hall, lacking altogether that potent charm of foregone association which no Murray can furnish.
>
> (Edwards 1877, 263)

And yet, while the best of the travel writers were supposed to invest the sights with profound meaning – Edwards's 'Poem' – they also participated in the same logic of anticipation and confirmation as the guidebook. 'After a hearty lunch,' one traveller wrote, 'we read Miss Martineau's beautiful and Sir Gardner Wilkinson's graphic descriptions of all we had seen' (A daughter of Japhet 1858, 184). Another described how 'we would read aloud Charles Dudley Warner's and Miss Edwards's opinions of our next stopping place' (Gibson 1899, 48). The textual attitude continued to structure and validate the experience of sightseeing in imaginative registers too.

VIEW-HUNTING IN THE CABINET OF CURIOSITIES

To some writers haunting tombs was an odd way to spend a holiday. George Ade called it 'the cemetery circuit' (1906, 185) and Kipling described Egypt as 'one big undertaker's emporium' (1920, 185) (Figure 4). Most tourists were drawn not only to the mausoleums and monuments, however, but also to the everyday life of the streets and bazaars in Cairo and the villages along the Upper Nile. To some the two were intimately connected. 'While you are doing tombs you are not spending a dark and gloomy day with the dead,' Lorimer insisted. 'You are transplanted into the midst of a very real and living life – the life of a people whose history has been written for us on the walls of their tombs and temples – [and which is] hardly different from the life of the people you see around you in Egypt today' (1909, 151). Arguments like this were not uncommon; by representing Egypt as an anachronistic space in which past and present existed *outside* the space of the modern, they claimed to open an imaginative (and extraordinarily presumptuous) passage into an ancient land.

But to many other tourists 'monumental Egypt' and 'living Egypt' were worlds apart. There were of course travellers who were utterly uninterested in the everyday lives of ordinary Egyptians, and who lost themselves in the contemplation of tombs and temples. Nightingale for one had no time for present-day Egypt, and as she sailed up the Nile she marvelled at how she and her companions had 'become so completely inhabitants of another age': a privilege from which she explicitly excluded local people (Nightingale 1854, 64).[16] But others surprised themselves. Cobbe confessed that she had always longed to see the monuments and that she 'had in fact hardly thought of any interest beyond the antiquities' – until she arrived in Cairo. Once there, however, she found 'the *living* interest around was so vivid, so intense, that I did not even desire to leave it for a day, and cheerfully deferred my visit to the Pyramids for more than a fortnight' (Cobbe 1864, 55–6). By the turn of the century many travel writers shared Kelly's suspicion that 'most of the Nile visitors are secretly bored, and only "do" these sights under moral compulsion' (Kelly 1904, 77). This was Ade's view too. While 'the guidebooks talk about rock tombs and mosques', he was convinced that most tourists 'find their real enjoyment in the bazaars and along the crowded streets [of Cairo] and on the sheer banks of the Nile which stand out as an animated

Figure 4 Edwardian tourists at lunch in a tomb.

panorama for hundreds of miles' (Ade 1906, 148). So was Sidney Low. Writing in 1914, he doubted 'that the modern tourist, as a general rule, takes the antiquities too seriously'. Most 'provide themselves with the volumes of Baedeker, Murray, or Flinders Petrie, and begin with an honest endeavour to assimilate those improving works.' But after a while their interest waned: they were, after all, 'in holiday mood, entirely resolved to enjoy themselves'. In consequence, he concluded,

Of modern Egypt – the real, living Egypt – they know even less than they do
of that ancient Egypt which still lies half buried under the dust; but the Egypt
of Messrs Cook, the Egypt of the hotels and the palace steamers, the Egypt of
the dragoman and the donkey-boy, the Egypt which dines and dances and
holds gymkhanas, the Egypt which enables the Northern sojourner to bask
and play in the sun – that they most keenly appreciate.

(Low 1914, 144)

So they did, but during the second half of the nineteenth century many of them
also became increasingly preoccupied with 'view-hunting', with seeking out an
exotic Orient that was far removed from both the tombs and temples and the
hotels and steamers. In 1845 Bartlett had considered Cairo 'a peculiar attraction'
for travellers who wanted to do more than merely 'despatch the "sights"'; but as
the years went by those 'peculiar attractions' were converted *into* the 'sights'
(Bartlett 1849, 61). For much of the period guidebooks played little or no part in
scripting these view-hunting expeditions because they were still largely concerned
with antiquities. Murray was 'so taken up with what happened 3000 years ago,'
complained one writer, 'so full of corrections of the assertions of other antiquar-
ies, that he can find scant room to deal with subjects of present interest' (Eden
1871, 123). Tourists therefore had to find the cabinet of curiosities and inspect its
contents through other accounts.

The most popular way of doing so was to enframe Cairo through the illuminated screen of the *Arabian Nights*. Here, for example, is Curtis:

Abon Hassan sat at the city gate, and I saw Haroun Alrashid quietly coming
up in that disguise of a Moussoul merchant. I could not but wink at Abon,
for I knew him so long ago in the *Arabian Nights*. But he stared rather than
saluted, as friends may, in a masquerade. There was Sinbad the porter, too,
hurrying to Sinbad the Sailor.

(Curtis 1858, 2–3)

He was right to suppose that few travel writers could get by without 'a dash of the
Arabian Nights'. The Reverend Smith thought it wonderful that this 'delight of
our childhood' – in its bowdlerized form it was a classic of the Victorian nursery –
literally materialized as he and his party ventured into the old city (Smith 1868a,
50), and writer after writer drew on its familiar characters to fill the streets of
Cairo and illuminate them like a magic-lantern show. The *Arabian Nights* occupied a central place in this Orientalist imaginary because it conjured up a fantasy
and yet at the same time – and in the same *place*, which was the point – its reality
seemed to be confirmed by the experience of riding through the narrow streets
and peeping into the crowded bazaars. Thus Bartlett described how 'creatures,
which once [seemed] so fanciful and visionary, seem to kindle into life and reality
as we gaze upon every object that surrounds us' (Bartlett 1849, 46). Cairo
became what he called 'a flittering phantasmagoria', half-illusion, half-reality
(Ibid., 55). And sixty years later Reynolds-Ball was still describing Cairo as a

street-theatre showing 'a living kinetoscope' of 'the *dramatis personae* of the *Arabian Nights*' (1907, 74–5). To many travel writers the *Arabian Nights* was essential reading for *both* 'the truth *and* poetry through which we see Eastern life as through some coloured and glowing medium' (Appleton 1876, 246).

By the end of the century, however, this imagery had become so hackneyed that Moberley Bell could satirize its hold by reciting Curtis's description (above) to his travelling companion as though it were his own. The reaction was instructive: 'The words sounded familiar to him. "There's something like that in an American book," he said. "Good heavens! You've read it then! Why, I sat up all last night learning pages on purpose to please you; and there's lots more of it ...."' (1888, 14–15). The point of the (I presume fictitious) anecdote is not so much that Bell was found out – Curtis's book was, after all, well known and much loved – but that his companion was disoriented by the description: 'In the dirty, semi-European faces around him,' Bell revealed, 'he could recognise none of the wonders described by his ordinarily practical friend' (Ibid.).

This did nothing to diminish the imaginative power of the *Arabian Nights*, but cultivated tourists did seek out more 'practical' – which is to say more ethnographic – accounts. The equivalent of Wilkinson's *Manners and Customs of the Ancient Egyptians* was Lane's *Manners and Customs of the Modern Egyptians*. First published in 1836, it was as 'indispensable' to Beaufort in 1861 (479) as it was to Appleton in 1876 (246), and into the first decade of the twentieth century tourists like Lorimer continued to watch 'the street life, its manners and customs through Lane's spectacles' (1909, 101). Lane went on to provide a new English translation of the *Arabian Nights*, but it was the ethnographic galleries of *Manners and Customs* that provided the model for other travel writers. They transformed Cairo into an open-air gallery. 'Never has a museum of human types offered a collection so varied, so complete, so picturesque,' declared one visitor (Charmes 1883, 78). 'It takes two or three days to rid oneself of the idea that the streets are parading their colour and movement and their endless variety of Oriental types and costumes for your diversion only,' wrote another, 'on an openair stage' (Butler 1909, 33). Many travel writers set these scenes in motion, taking their tourist-readers through the streets and bazaars by parading through their pages a dazzling succession of subordinate clauses, each one displaying a different 'type'. Even Edwards succumbed:

> Here are Syrian dragomans in baggy trousers and braided jackets; barefooted Egyptian fellaheen in ragged blue shirts and felt skull-caps; Greeks in absurdly stiff white tunics, like walking penwipers; ... swarthy Bedouins in flowing garments; ... Englishmen in palm-leaf hats and knickerbockers, dangling their long legs across almost invisible donkeys; native women of the poorer class, in black veils that leave only the eyes uncovered . . .; dervishes in patchwork coats. . . .
>
> Now a water-carrier goes by, bending under the weight of his newly-replenished goatskin. . . . Now comes a sweetmeat-vendor . . .; and now an Egyptian lady on a large grey donkey led by a servant. . . . Next passes an

open barouche full of laughing Englishwomen; or a grave provincial sheikh all in black, riding a handsome Arab bay . . .; or an Egyptian gentleman in European dress. . . . Next passes an itinerant lemonade seller, with his tin jar in one hand, and his decanter and brass cups in the other; or an itinerant slipper-vendor with a bunch of red and yellow morocco shoes dangling on the end of a long pole. . . .

(Edwards 1877, 4–6)

And on and on. Apart from the intrusive encroachments of Englishmen, Englishwomen and the 'Egyptian gentleman in European dress', passages like this recur in book after book. But it was precisely the presence of those European actors that placed such a premium on 'view-hunting', that valorized sightseeing as the successful capture of an exotic, supposedly 'timeless' and hence 'authentic' Orient which remained 'unspoiled' by the invasions of modernity – including the whole apparatus of tourism that had turned the old city and its inhabitants into sights.

Yet in 1873 Murray was still content merely to quote Lane, one authority reinscribing another, whereas his rivals started to mimic *Manners and Customs* by providing their own ethnographic galleries. Baedeker was the most supremely self-confident. He awarded Cairo's 'street scenes' two asterisks – his highest accolade and one which Cobbe would have been pleased to learn placed them on the same level as the Pyramids – because he said they 'afford[ed] an inexhaustible fund of amusement and delight, admirably illustrating the whole world of Oriental fiction' (Baedeker 1898, 33). That last clause reinscribes the master trope: the street scenes were at once literally fantastic, unreal, belonging to the world of the *Arabian Nights*, and yet also capable of validation – of being *shown* by being *there* – so that, as Curtis put it, 'the romance of travel [was] real' (1858, 7). Baedeker conducted the tourist-reader on an Orientalist *flânerie*, in which the old city became a 'magic labyrinth' and Baedeker its curator-guide. Here he is, plunging into the narrow streets off the Muski with single-minded determination:

[We] diverge by the Shâr'ia Hammâ et-Talât to the right, and follow the first lane to the left . . . passing a red and yellow mosque on the right, and disregarding the attractions of the European glass wares sold here. Pursuing a straight direction (i.e. as straight as the crooked lanes admit of), we pass an Arabian gateway on the left, and, on the right, the end of a narrow lane, through which we perceive the entrance to the uninteresting Greek Orthodox Church.

Farther on, beyond the covered entrance of a bazaar in ruins, we turn once more to the right by the Shâr'ia el-Hamzâwî es-Saghîr, in which is the bazaar of the same name.

(Baedeker 1898, 41)

Having arrived at the site, Baedeker directed the sight. It was only when 'the traveller has learned to distinguish the various individuals who throng the streets, and knows their different pursuits, that he can thoroughly appreciate his walks,' Baedeker advised, so he offered 'a brief description' – an illustrated programme – of 'the leading characteristics of different members of the community' (Ibid., 34–8) (Figure 5). Such stereotypes were sharply criticized by one writer, who objected that 'the people, though clothed in Eastern garb, are not of one type or mould, but are good-humoured, hard-working *individuals*, with many of whom the traveller will in time become acquainted' (Kelly 1904, 3). But such impassioned observations were very much the exception. For most writers it was precisely the costume – 'Eastern garb' – that scripted the scene as theatre and constituted local people as actors and tourists as audience.

The projection of the *Arabian Nights* and the parade of ethnographic types fed off one another and turned Cairo into an entertainment as well as an

Figure 5 Baedeker's 'water-carrier'.

exhibition; together they domesticated its exoticism, making it both attractive and innocuous. But fairy tales always have their dark sides, and by the turn of the century – as modernity closed in – many travel writers sought to provide a sense of intrigue, even danger, that went far beyond the cautionary observations made by earlier travellers. This was achieved in several ways. If guidebooks advised tourists where they *ought* to go, other travel writers made much of the places where they ought *not* to go. In the middle decades of the nineteenth century many Nile travellers had been entranced (or repulsed) by performances of female singers and dancers – the *Ghawazi* – who had been exiled from Cairo to Kena, Esne and Assouan for their supposedly licentious behaviour. Flaubert's night with Kuchuk Hanem is perhaps the most well known and certainly the most intimate account, but many other travellers reported less erotic but still affecting experiences at these *fantasia* (see Steegmuller 1979, 112–19). By the closing decades of the century, however, such performances had become shadows of their former selves, and the trade had returned to Cairo. Many travel writers warned tourists of 'the Fishmarket', the quarter 'where vice reigns and flourishes' and where for a few francs 'one can assist at scenes of the most revolting immorality'. In 1893 de Guerville confessed that he had made 'rather a risky arrangement' with a donkey-boy to take him to one of the bawdy houses. Seated on a stool in a dimly lit room he was at first 'disgusted and rather scared', but when the belly-dancers appeared, 'with heads and breasts now thrown back, now thrown forward to within a hand's-breadth of my face, their flesh quivering, the scent of their bodies in the air, their harsh cries joined to wild music,' he was, he said, 'completely overcome' (de Guerville 1906, 78–9). Sladen was made of sterner stuff. For the curious but faint of heart, he advised taking a cab through the Fishmarket during the day, when 'the strange women of the Bible, lolling about in sufficient numbers, give some idea of the place to those who could not endure its shamelessness at night' (Sladen 1911, 22–3). By night, however, it was 'a blaze of Oriental vice':

> The women . . . positively flame with crimson paint and brass jewellery and have eyes flashing with every kind of mineral decoration and stimulant, and far too much flesh. If you walk through the Fishmarket when they are prowling for victims, your clothes are nearly torn off in the agonised attempts to secure your attention.
>
> (Sladen 1911, 61)

Sladen's account, and others like it, served to establish the travel writer as a man of the world, the sort of guide who could be relied upon because he had been everywhere (though presumably not done everything), and also to conjure up the usual masculinist fantasy of the Orient as a liminal zone of unrestrained sexuality. It was figured not only as a space of debauchery but as a space of availability, its denizens reaching out, 'in the agonised attempts to secure your attention', to threaten the integrity and dignity – the very sovereignty – of the tourist, turning him from spectator to voyeur, and perhaps worse.

Other writers tried to achieve a sense of danger through an equally stylized engagement with Islam, which was invariably – ignorantly and insultingly – referred to as 'Mohammedanism' (Said 1979, 60, 66). To some writers the mosques themselves were merely the stage, and not always very interesting at that: 'When you have seen one,' sighed a weary Eames, 'you have seen them all' (Eames 1855, 49). In the middle decades of the century entering a mosque had been a theatrical production in its own right. It was necessary to adopt 'Turkish dress,' Wilkinson advised, 'and though the wearer be known to be an European, the adoption of a Moslem costume suffices to prevent objection' (1843, 229). But by the end of the century it was much more prosaic: 'The mosque now being a "sight" more than a place of worship,' Tyndale observed, 'a fee is charged for admittance' (Tyndale 1912, 25). Yet it was still possible for writers to tantalize the tourist-reader by moving from fantasy – Islam as 'the religion of the *Arabian Nights*', as one writer put it – to fanaticism. Visiting the mosque of el-Azhar, for example, Baedeker warned that this was 'one of the fountain-heads of Mohammedan fanaticism' so that 'the traveller should, of course, throughout his visit be careful not to indulge openly in any gestures of amusement or contempt' (1898, 45). The advice was echoed by Reynolds-Ball. 'The authorities do not altogether encourage the presence of strangers,' he wrote (1907, 56); some of the sects 'are decidedly fanatical, and strangers will be well advised to abstain from any overt expression of amusement at the extraordinary spectacle' (Ibid., 61). In advising their readers against betraying their reactions, these authors were clearly sanctioning them: amusement or contempt may have been dangerous but they were also construed as perfectly *proper* responses. The same writers advised still greater caution during a visit to the Dervishes. Here is Reynolds-Ball again:

> The religious exercises of these fanatical orders are decidedly repulsive spectacles, but as they are among the recognized sights of Cairo – in fact to enable strangers to witness the spectacle (which usually begins at 2 P.M.) the table d'hôte lunch at the chief hotels takes place an hour earlier than usual – it is necessary to notice them.
>
> (Reynolds-Ball 1907, 113)

There were several orders of dervishes, but *fin-de-siècle* travel writers were most exercised by the Howling Dervishes. There was, according to Reynolds-Ball, 'a certain element of genuine fanaticism in the performance when at its height that might prove dangerous to the spectators'.

> Ladies are not advised to remain to the end; or, if the spectacle proves too engrossing, they should be especially careful not to sit to close to the dervishes, or to brush against the performers [who] maintain that the touch of a woman is contamination. . . . Male visitors, too, will be well advised to avoid letting it be seen that they are affected by the ludicrous aspect of some phases of this performance.
>
> (Reynolds-Ball 1907, 194–5)

The reproduction and circulation of these stereotypes was accelerated by the development of photography. Studio portraits and postcards of ethnographic 'types' were available for sale, and there seems to have been a lively trade in pornographic images too, but it was the rise of amateur photography that spurred view-hunting at the end of the century. Sladen's own *Oriental Cairo: The city of the Arabian Nights* (1911) showed 'explicitly where each kind of unspoiled native life is to be found' (11), and Sladen revelled in identifying 'poor Egyptians with the most kodakable attitudes and occupations' (35). He directed his readers to 'unspoiled native streets' because, even though 'there is nothing for the European to buy in the shops' – 'they cater for humble native wants' – 'there is plenty for him to photograph among the shoppers' (83). In such places, he observed, there was hardly anything 'thought worthy of mention by Baedeker or Murray', yet they were 'paintable' and 'kodakable' from end to end (147).[17]

SCRIPTING EGYPT

'It is not what a country is,' Warburton had mused in 1844, 'but what we are that renders it rich in interest or pregnant with enjoyment' (90). This claims too much, but the production of imaginative geographies of Egypt by European and American travel writers during the long nineteenth century owed much to the fantasies and fears of modern Orientalism. The essays that accompanied the stunning plates of the Napoleonic *Description de l'Égypte* in the early decades of the nineteenth century constructed a fantasy land of a rationalist Egypt, for example, 'a sort of Eden where reason triumphed', 'a pefect world' that appeared to foreshadow the ideals of revolutionary France with an astonishing fidelity (Laurens *et al.* 1989, 352–3). So too, as the century wore on, the production of itineraries and the conversion of 'sites' into 'sights' was also about the scripting of 'the West's Egypt': a project – and an act of possession – sardonically advertised by Gibson towards the end of the century when he advised his readers en route from Alexandria 'to look out of the right-hand window of the [railway] car for a first glimpse of the Pyramids, the first sure proof that you are in *Napoleon's* Egypt' (Gibson 1899, 9; my emphasis).

But this Egypt was not all in the mind; it had its own dense materiality. The travellers and tourists I have followed in these pages did go to Egypt, and, so far as one can tell, they did act in the ways outlined here; *dahabeeahs* and dragomans were hired, hotels and tourist-steamers were built, guidebooks and travel accounts were published and put to use. It is necessary to insist on all this because so many of Said's critics want to know whether Orientalism was more than a distorting mirror: is it possible, they ask, to get behind these images and disclose the 'real' Egypt? The Egypt of these imaginative geographies was constructed with little or no reference to those who lived there, and no doubt its assumptions and appropriations provoked both distrust and dissent: but it was none the less a 'real' Egypt too, and its productions had real consequences. In the first place, these scriptings were intimately involved in the fabrication of a space of constructed visibility: in the production of modern vantage-points and viewing platforms, in opening up (and cleaning up) an 'ancient Egypt' and in creating and displaying an 'exotic Orient'. These achievements entered into the larger creation of what was heralded as a 'new Egypt', the Egypt of the Suez Canal, of 'Paris-on-the-Nile', of Thomas Cook & Son, and of British occupation, developments which involved the pursuit of European and American political and economic interests and the sustained complicity of successive Egyptian élites. In the second place, these scriptings were involved in the constitution of the modern tourist as spectator-voyeur, as consumer-collector and, above all, as sovereign-subject. All three were implicated in the powers and practices of colonialism, but it was the last figure that invited tourists to experience and command Egypt as a series of scenes set up for their own edification and entertainment. Scripting Egypt was always about much more than putting on a play.

NOTES

1 I have borrowed the basic idea from Sharon Marcus.

2 There were exceptional figures like Edward Lane in Cairo or Lady Gordon at Luxor who lived among the local people, adopted many of their customs and learned Arabic, often deliberately setting themselves apart from other 'Franks'. In paying homage to them, however, other travellers constantly drew attention to their unusual position and so confirmed the norm. Thus Curtis described Lane as 'the eastern Englishman' (Curtis 1858, 4), and when Warren paid a call on Lady Gordon he found 'an eccentric woman . . . dressed in half-Arab costume' who received him 'with as much sang froid as an Eastern princess' (Warren 1883, 54).

3 The discursive practices of 'belated Orientalism' are 'inscribed within both the economies of colonial power and the exoticist desire for a disappearing Other' (Behdad 1994, 14).

4 One writer attached a list of recommended dragomans, but warned that 'as their characters are continually changing . . . their last certificate should be examined and inquiries made of the Consul' (Hoskins 1863, 93).

5 'It is but seldom that ordinary travellers can have any direct communication with the people of the country' (Manning 1875, 97). One of the classic (comic) vignettes of the attempt to carry on a conversation through an intermediary is the dialogue between a traveller, his dragoman and a Pasha staged by A.W. Kinglake in his *Eothen* (1928, 14–18). Glossaries and phrase-books were presumably of some help, though many of them were preoccupied with the language of complaint and command. 'With commendable energy travellers frequently, and after much labour, learn a few Arabic phrases, usually questions, forgetting that they cannot possibly understand the replies,' observed one writer. 'Let me recommend them to confine their earlier efforts to the acquisition of such sentences only as give absolute instructions to servants, drivers, etc., and to which no response is required' (see Kelly 1904, 13–14). It is difficult to know the extent to which tourists put glossaries and phrase-books to use, but at the very least they provided a horizon of expectation – another scripting – within which interaction could be conducted.

6 Cf. Belzoni (1821, 83).

7 A local sheikh or even a district governor was often enlisted to punish miscreants – usually with the *bastinado* – but in doing so travellers were clearly demonstrating

that their power reached far up the local chain of command. In such circumstances they expected to receive what Smith called 'the customary acquiescence in European demands' (though in this particular case, ironically, he and his party were the exception that proved the rule: they were unable to prevail over the governor) (Smith, 1868b, 102).

8 While it is difficult for me to recover the reactions of local people to these productions with any degree of completeness (and not only because I don't speak Arabic; it is by no means clear that an archive of comparable richness survives, or even existed, adequate to the responses of ordinary Egyptians), their voices are by no means absent from travellers' accounts. It is of course the case that they are always highly mediated: the tourists give their version of events, their interpretations of the practices of shopkeepers in the bazaars, dragomans and donkey-boys, and villagers; on occasion they report what is supposedly direct speech (though since few of them spoke more than a few words of Arabic and most of the locals had a limited if serviceable command of English, those exchanges were almost always oblique and of necessity constrained). Still, it is possible to read European and American reports of these encounters 'against the grain', as it were, and to glimpse spaces in which local people retained their independence and integrity in various ways. This isn't to rehearse yet another example of unremarked but nonetheless 'heroic' resistance to the colonial usages of foreign tourists and travellers, but it is to open up the possibility of a more foliated space of transculturation than would otherwise be visible. Since this raises a series of demanding – and vitally important – methodological questions, I address the issue in detail in another essay.

9 In the first half of the nineteenth century flying a national flag was a concession rather than a custom. As one American explained, 'It is necessary here for every stranger to place himself under the flag of his country, else his boat and men are liable to be taken at any moment by officers of the pasha.' But even then it also signified sovereignty: 'It was the first time I had myself ever raised the banner of my country,' he confessed, 'and I felt a peculiar pride in the consciousness that it could protect me so far from home' (Stephens 1837, 40). Confusion had set in by the early 1840s: Wilkinson (1843, 87, 214–15) announced that 'the privilege accorded to travellers of hoisting their flag has been voluntarily renounced by the English; but some still do this, as well as travellers of other European nations'. This state of affairs provoked a number of ugly incidents between tourists and troops, and as late as the 1850s it was still necessary to fly the tourist's own national flag to protect the crew from conscription (see Eames, 1855, 55).

10 There was, of course, quite another 'society of the River'. 'When dahabeahs meet,' Appleton noted, 'there is often much racing about of sandals [skiffs] and little boats, as this whole fraternity of dragomen, waiters and cooks is but one great Corporation of the Nile' (1876, 217).

11 Stephens also gloried in the freedom: 'Think of not shaving for two months, of washing your shirts in the Nile and wearing them without being ironed. True, these things are not absolutely necessary, but who would go to Egypt to travel as he does in Europe? "Away with all fantasies and fetters" is the motto of the tourist. We threw aside pretty much everything except our pantaloons' (1837, 35). Similarly, Prime freely confessed to 'eschew[ing] all manners of dress' once up river: 'It would be impossible to say what style of national costume I wore.' He was usually barefoot, and wore only a thin pair of linen trousers, a blue shirt and tarbouche (1857, 183). During the same period, in contrast, many women travellers attached great importance not only to appropriate dress but also to its proper ironing (e.g. Martineau 1848, 72; Eames 1855, 68–9).

12 The verdict was, perhaps, unduly harsh. Martineau was certainly more sympathetic than many tourists towards her crew: 'We do not agree with travellers who declare it necessary to treat these people with coldness and severity – to repel and beat

them.' But she immediately enlisted one of the standard tropes of colonial discourse as a counter-example: 'We treated them as children; and this answered perfectly well' (Martineau 1848, 40). This may also explain her amusement at the ways in which the crew mimicked the behaviour of their employers (Ibid., 36–7).

13 She gave two reasons: 'We are not worthy yet of this great unveiling: and the inhabitants are not, from their ignorance, trustworthy as spectators.'

14 In any case the romance of the ruins was always problematic in a land where sun and sand conspired to preserve them so well. Bartlett noted 'the absence of those grey hues and weather stains and that overgrowth of vegetation which give so venerable an air to those of Europe' (1849, 179), while Frederic Eden lamented that the sharpness and precision of the monuments 'destroys the beauty of ruined architecture' (1871, 51). It was in order to recuperate the romantic gaze that moonlight visits to the Pyramids and to Karnak became fashionable.

15 These practices continued throughout the nineteenth century. In 1891 Reeve, travelling upriver on Thomas Cook's *Mohammed Ali*, complained that 'there was supposed to be an ample supply of stationery on board, but certified copies of twenty-seven ponderous diaries, of doubtful interest to anybody except the writers, exhausted the supply before we had gone two hundred miles' (1891, 75). Steevens described a similar voyage on *Rameses the Great* in 1898. As they approached their first stop, there was great excitement: 'our trusty cameras were slung at our backs: our diaries lay in our cabins with our stylo-graphs at half-cock beside them' (1898, 205).

16 See also Gregory (1995).

17 For a detailed discussion, see Gregory (1999).

REFERENCES

Ade, G. (1906) *In Pastures New*, New York: McClure, Phillips.

Appleton, T.G. (1876) *A Nile Journey*, Boston: Roberts.

Arnold, J. (1882) *Palms and Temples: Four months' voyage upon the Nile*, London: Tinsley Brothers.

Bacon, L. (1902) *Our Houseboat on the Nile*, Boston: Houghton and Mifflin.

Baedeker, K. (1898) *Egypt: Handbook for travellers*, Leipzig: Baedeker.

Bartlett, W.H. (1849) *The Nile Boat, or glimpses of the land of Egypt*, London: Hall, Virtue.

Beaufort, E. (1861) *Egyptian Sepulchres and Syrian Shrines*, vol. 1, London: Longman, Green.

Behdad, A. (1994) *Belated Travelers: Orientalism in the age of colonial dissolution*, Durham, NC: Duke University Press.

Bell, C.F.M. (1888) *From Pharaoh to Fellah*, London: Wells Gardner, Darton.

Belzoni, G. (1821) *Narrative of the Operations and Recent Discoveries within the Pyramids, Temples, Tombs and Excavations in Egypt and Nubia*, 2nd edition, London: John Murray.

Bredon, P. (1991) *Thomas Cook: 150 years of popular tourism*, London: Secker and Warburg.

Butler, E. (1909) *From Sketch-Book and Diary*, London: Adam and Charles Black.

Buzard, J. (1993) *The Beaten Track: European tourism, literature and the ways to 'culture' 1800–1918*, Oxford: The Clarendon Press.

Carey, M.L.M. (1863) *Four Months in a Dahabeeh, or Narrative of a winter's cruise on the Nile*, London: Booth.

Charmes, G. (1883) *Five Months at Cairo and in Lower Egypt*, trans. William Conn, London: Richard Bentley.

Cobbe, F.P. (1864) *The Cities of the Past*, London: Trübner.

Curtis, G.W. (1858) *Nile Notes of a Howadji*, New York: Harper.

Curtis, W. (1905) *Egypt, Burma and British Malaya*, Chicago: Revell.

Curzon, R. (1849) *Visit to Monasteries on the Levant*, London: John Murray.

A daughter of Japhet (1858) *Wanderings in the Land of Ham*, London: Longman, Brown.

de Guerville, A.D. (1906) *Modern Egypt*, London: Heinemann.

Dunning, H. (1905 [1907, 1909]) *To-Day on the Nile*, New York: James Potts.

Eames, J.A. (1855) *Another Budget, or Things which I saw in the East*, Boston: Ticknor and Fields.

Ebers, G. (1879) *Egypt: Descriptive, historical, picturesque*, vol. 2, trans. Clara Bell, London: Cassell.

Eden, F. (1871) *The Nile without a Dragoman*, London: King.

Edwards, A. (1877) *A Thousand Miles up the Nile*, London: Longmans Green.

Ferguson, R. (1873) *Moss Gathered by a Rolling Stone*, Carlisle: Thurnam.

Flinders Petrie, W.M. (1892) *Ten Years' Digging in Egypt*, London: Religious Tract Society; New York: F.H. Revell.

Gibson, C.D. (1899) *Sketches in Egypt*, New York: Doubleday and McClure.

Gordon, Lady Duff (1969) *Letters from Egypt 1862–1869*, ed. Gordon Waterfield, New York: Praeger.

Gregory, D. (1995) 'Between the book and the lamp: imaginative geographies of Egypt, 1849–50', *Transactions of the Institute of British Geographers* 20, pp. 29–57.

—— (1998) 'Emperors of the gaze: photographic practices and the captivation of space in Egypt, *c.* 1839–*c.* 1914', in J. Ryan and J. Schwartz (eds), *Picturing Place: Photography and imaginative geographies*, London: John Wiley.

Grewal, I. (1996) 'The guidebook and the museum', in *Home and Harem: Nation, empire and the cultures of travel*, Durham, NC: Duke University Press, pp. 85–130.

A Handbook for Travellers in Egypt (1873) London: John Murray.

Hopley, H. (1869) *Under Egyptian Palms, or Three bachelors' journeyings on the Nile*, London: Chapman and Hall.

Hoskins, G.A. (1863) *A Winter in Upper and Lower Egypt*, London: Hurst and Blackett.

Kelly, R.T. (1904) *Egypt Painted and Described*, London: Black.

Kinglake, A.W. (1844 [1928]) *Eothen*, Oxford: The Clarendon Press.

Kipling, R. (1920) *Letters of Travel*, London: Macmillan.

Lane, E.W. (1836) *Manners and Customs of the Modern Egyptians*, London: Charles Knight.

Laporte, L. (1872) *Sailing on the Nile*, trans. Virginia Vaughan, Boston: Roberts.

Laurens, H., Gillispie, C., Golvin, C. and Traunecker, C. (1989) *L'Expédition d'Égypte 1798–1801*, Paris: Armand Colin.

Leland, C. (1875) *The Egyptian Sketch-Book*, New York: Hurd and Houghton.

Lorimer, N. (1909) *By the Waters of Egypt*, London: Methuen.

Low, S. (1914) *Egypt in Transition*, London: Smith, Elder.

Manning, S. (1875) *The Land of the Pharaohs: Egypt and Sinai illustrated by pen and pencil*, London: Religious Tract Society.

Marden, P.S. (1912) *Egyptian Days*, Boston and New York: Houghton Mifflin.

Martineau, H. (1848) *Eastern Life, Present and Past*, Philadelphia: Lea and Blanchard.

Maule, F. (1901) *Only Letters*, Philadelphia: George Jacobs.

Mitchell, T. (1988) *Colonising Egypt*, Cambridge: Cambridge University Press.

Montbard, G. (1894) *The Land of the Sphinx*, London: Hutchinson.

Nightingale, F. (1854) *Letters from Egypt*, London: Eyre and Spottiswoode. Reprinted as *Letters from Egypt: A journey on the Nile 1849–1850*, ed. Anthony Sattin, New York: Weidenfeld and Nicolson.

Penfield, F.C. (1903) *Present-Day Egypt*, New York: Century.

Prime, W. (1857) *Boat Life in Egypt and Nubia*, New York: Harper.

Reeve, C.M. (1891) *How We Went and What We Saw*, New York: Putnam.

Reynolds-Ball, E. (1907) *Cairo of To-Day*, London: Adam and Charles Black.

Roberts, E. (1841) *Notes of an Overland Journey through France and Egypt to Bombay*, London: W.H. Allen.

Said, E. (1979) *Orientalism*, London: Penguin.

Sladen, D. (1908) *Egypt and the English*, London: Hurst and Blackett.

—— (1910) *Queer Things about Egypt*, London: Hurst and Blackett.

—— (1911) *Oriental Cairo: The city of the 'Arabian Nights'*, London: Hurst and Blackett.

Smith, A.C. (1868a) *The Attractions of the Nile and its Banks*, vol. 1, London: John Murray.

—— (1868b) *The Attractions of the Nile and its Banks*, vol. 2, London: John Murray.

Steegmuller, F. (ed.) (1979) *Flaubert in Egypt: A sensibility on tour*, Chicago: Academy.

Steevens, G.W. (1898) *Egypt in 1898*, London: Blackwood.

Stephens, J.L. (1837) *Incidents of Travel in Egypt, Arabia Petrae and the Holy Land*, New York: Harper.

Taylor, B. (1854) *A Journey to Central Africa, or Life and landscapes from Egypt to the Negro kingdoms of the White Nile*, New York: Putnam.

Thackeray, W. (1844 [1991]) *Notes on a Journey from Cornhill to Grand Cairo*, Heathfield: Cockbird Press.

Tirard, H. and Tirard, N. (1891) *Sketches from a Nile Steamer*, London: Kegan Paul.

Twain, M. [Samuel Clemens] (1869) *The Innocents Abroad*, New York: American Publishing Co.

Tyndale, W. (1912) *An Artist in Egypt*, London and New York: Hodder and Stoughton.

Wallis Budge, E.A. (1906) *Cook's Handbook for Egypt and the Sudan*, 2nd edition, London: Thomas Cook.

Warburton, E. (1844 [1855]) *The Crescent and the Cross, or Romance and realities of Eastern travel*, London: Colburn.

Ward, J. (1900) *Pyramids and Progress*, London: Eyre and Spottiswoode.

Warner, C.D. (1876) *Mummies and Moslems*, Toronto: Belford.

Warren, W.W. (1883) *Life on the Nile in a Dahabeeh, and Excursions on shore between Cairo and Assouan*, 3rd edition, Boston: Lee and Shephard.

Wilkinson, J.G. (1837) *Manners and Customs of the Ancient Egyptians*, vol. 3, London: John Murray.

—— (1843) *Modern Egypt and Thebes*, London: John Murray.

Withey, L. (1997) *Grand Tours and Cook's Tours: A history of leisure travel, 1750–1915*, New York: William Morrow.

7 Dis-Orientation

On the shock of the familiar in a far-away place

James Duncan

Victorian travellers from the 1840s on described the Kandyan Highlands of Ceylon as one of the most beautiful places in the world. Having lived there for a year and finding it very beautiful myself, I nevertheless wondered why the Victorians found it quite so lovely. The answer, I believe, lies in the region's hybridity, for these travellers' accounts operated through a set of exoticizing and familiarizing gestures. Victorian writers, men and women alike, were shocked simultaneously by the uncanny familiarity of the place, and by its alterity. And yet this shock was domesticated, which is to say turned into delight, by a textualized way of seeing based upon a form of hybridity which did not evolve on this spot, but rather was invented half a century before, and 7,000 miles away in Britain.

This study then explores the coming together of a particular imaginative geography, in this case a romantic one, with a particular place, the Kandyan Highlands. Until 1815 when the Kandyan Kingdom was conquered by the British, it was the site of another imaginative geography, that of the heaven on earth of the Buddhist god-king.[1] This Kandyan imaginary was impressed on the landscape – on sacred mountains, rivers, town layouts, palaces and temples. It was encoded in the ritual and secular practices of the Kandyans, whose politics were no less real for being imagined. This imaginary geography, in other words, was inseparable from the concreteness of this place. In fact the very specificity of this place was in large part produced by the impress of this imaginary. After 1815, this particular imaginative geography became increasingly difficult for the Kandyan people to sustain as it became manifestly clear that the new British rulers intended to remain; for new political imaginations produce new imaginary geographies. At times running beside this Kandyan imaginative geography, at times running roughshod over it, was a British imaginary which reworked the former as the picturesque. This new imaginary was a form of translation which recuperated the Highlands for a British audience. In doing so it not only re-imaged the Kandyan imaginary, forcing the politics of the Buddhist god-king through the taxonomic grid of romanticism and utilitarianism, but re-imagined the physical place itself, the mountains and the lakes and forests. What this complex, unstable translation of the culture and nature produced was a hybrid creation, which began as a discourse, a way of seeing and talking about a place, and ended as a reconstruction of that place, as a concretization of that new imagined geography.

In this paper I will focus on the way European travellers in the nineteenth century translated the Kandyan Highlands; how in particular the picturesque was the trope through which this landscape was read (Cardinal 1997). The particular imaginative geography that the British chose was influenced by the physical nature of the Kandyan Highlands themselves. I will then go on to show how this way of seeing, shaped by the landscape, was further inscribed on the landscape, thereby reinforcing future readings along the well-worn channels of meaning established by romanticism. It was this tacking back and forth between the imaginary and the concreteness of the Highlands which produced the specificity of this place for the British in the nineteenth century. Finally I will briefly go on to demonstrate the ambivalence which this romantic reading of the Highlands produced for British settlers there. This place looked uncannily like highland Britain and therefore was a mirror of home. But it was not home. The discursive frame of the picturesque which placed an emphasis upon savagery, wildness and in particular decay simultaneously served as an aesthetic ideal and as a source of anxiety for settlers. For the dark side of the picturesque was decay figured as environmentalism and cultural degradation. Here the discursive frame of the picturesque escaped the aesthetic and the touristic and entered daily life. The British who were residents in, rather than visitors to, this picturesque place feared that they were part of the cultural decay of the place. They could not unambivalently maintain the distanced aestheticized view of the tourist, for they were not on the outside looking in – they were part of the landscape itself.

By the middle of the eighteenth century in Britain travellers were increasingly viewing Britain through a romantic frame which privileged the picturesque, a 'pleasing melancholy', and through the sublime. While a sublime sensibility sought out wild, mountainous landscapes, the picturesque sought 'irregular forms in natural scenery and the works of man [*sic*]' (Andrews 1989, 45). Writers like Gilpin in his *Observations on the River Wye* (1782) suggested that, henceforth, travellers view the country by the rules of the picturesque. This entailed not only an aestheticizing vision but within it a focus upon the remote, the primitive, the humble and the ruined (Andrews 1989, 64). Such a romantic discursive frame was not to be restricted to Britain. Imperialism meant that it could be applied to new fields of vision.

By the mid- to late eighteenth century the British Nabobs who worked for the British East India Company were returning from India not only with fortunes but with certain Indian aesthetic ideas as well. This new 'Indian taste' was to be recuperated to the picturesque. In 1786 the eminent Sir Joshua Reynolds in his *Discourses* suggested that architects consider moving from the plain Grecian influence to 'Gothick' and 'the barbaric splendour of those Asiatic buildings' (Head 1986, 21). In this same year the upper classes were being introduced to the picturesqueness of the Indian subcontinent by traveller-painters such as Hodgeson (*Select Views 1786–88*). These were followed by the popular paintings of Thomas and William Daniels from 1795 on (*Oriental Scenery*). In the late eighteenth and early nineteenth centuries some country estates were beginning to be orientalized in the manner suggested by Reynolds. The most famous example of such architecture was the Brighton Pavilion commissioned by the Prince Regent and completed in 1821, but there were many others (Head 1986). By the late eighteenth century, therefore, not only had the Orient been recuperated through the lens of the picturesque, it had literally been brought home to Britain as a form of hybrid landscape. In this sense, Trinh Minh-Ha's (1989) oft-quoted description of the post-colonial world as 'the Third world in the First world; the First in the Third' was an early nineteenth-century phenomenon.

By the mid-nineteenth century, travellers to Ceylon saw it through the lens of the picturesque. They had already been taught to see it through writings and paintings before they arrived there. One of the most successful of these writers was Campbell (1843), who divided his account into a series of views and used each one 'as the starting point for the exercise of romantic fantasy' (Gooneratne 1968). The Kandyan Highlands, therefore, simultaneously drew upon not one, but two sources of the picturesque: the Oriental, with its romantic imagery of decline, decay and loss, which Edward Said (1979) has described as Orientalism; and alpine Europe, beloved of travellers in search of picturesque views.

At the beginning of British rule in 1815, the colonialist hybridity which travellers noted, in the capital at least, was largely produced out of economic necessity. In the early years the administration had little money and therefore colonial institutions were housed in Kandyan buildings that were replete with the symbolism of the god-king. For example, the royal palace became the residence of the government agent and other palace buildings were transformed into a European hospital and a library. The large houses of the nobles were converted into stores and hotels. Such hybridity was made all the more picturesque by the ruin into which much of the capital was allowed to fall during the first decades of British rule. Particularly picturesque were the temples in the capital as the British purposefully cut off funding in the hope that the power of the Buddhist monks would decline.

The mountainous topography of the area around the capital and the relatively cool climate, for a tropical location, lent themselves admirably to a peculiar type of cultural production, the re-creation of a bit of Britain in the tropics. But this wasn't just any vision of Britain, it was a romanticized and highly aestheticized portrait of pre-industrial England. Promenades were created on the outskirts of Kandy, the vegetation was pruned and benches were set up at strategic spots to reveal the best views of the town, lake and the surrounding mountains (Dougherty 1890, 102; Willis 1907). A church was built whose tower 'had creepers making their way up it and looking for all the world like one of our home village churches, peaceable and unobtrusive and quite venerable' (Gregory 1894, 281). Surrounding the lake are 'bungalows half hidden in their gardens' (Woolf 1914, 76), full of English flowers which could just manage to survive in Kandy. Just as parts of Britain were being Orientalized in these years, so parts of Ceylon were being Anglicized. The picturesque, therefore, is not simply a way of seeing, it is simultaneously a way of doing, a way of world-making.

While the area around the old capital of Kandy had been thoroughly inscribed with the ideology of the Buddhist god-king and, therefore, the British changes

Figure 1 Painting of Kandy, 1864, by Captain O'Brien.

could at best produce hybridity, the same could not be said of the higher-altitude, much more lightly inhabited parts of the Highlands. Here, in a landscape virtually untouched by the Kandyan imaginary, the British could create a more mimetic portrait of Britain than was possible in the mountains around Kandy. This was precisely what Sir Samuel Baker, who later went on to achieve fame exploring Africa, set out to do. Baker first saw the hill station of Nuwara Eliya in 1847 when he went there to convalesce. As it lay at 6,000 feet, the weather was much cooler than at the old capital of Kandy, which lay at 1,600 feet. Here in the high mountains, travellers could truly imagine they were back in Britain. Baker demonstrates the contradictory strands of Enlightenment rationality and romanticism which lie uneasily together in so many pieces of landscape description. He adopts the former style because he is no mere tourist. Rather he plans to live in this place. He therefore indulges in an imperialist, utilitarian reverie that is far removed from the picturesque:

> Why should this place lie idle? Why should this great tract of country in such a lovely climate be untenanted and uncultivated? How often I have stood upon the hills and asked myself this question when gazing over the wide extent of undulating forest and plain! How often I have thought of the thousands of starving wretches at home who here might earn a comfortable livelihood! And I have scanned the vast tract of country, and in my imagination I have cleared the dark forests, and substituted waving crops of corn, and peopled a hundred ideal cottages with a thriving peasantry.
>
> (Baker 1855, 11)

Baker's vision is of transforming the Kandyan Highlands into rural England with himself as its master.

> In my determination to reside at Newera Ellia, I hoped to be able to carry out some of those visionary plans for its improvement which I have before suggested; and I trusted to be able to effect such a change in the rough face of Nature in that locality as to render a residence at Newera Ellia something approaching a country life in England, with the advantage of the whole of Ceylon as my manor, and no expense of gamekeepers.
>
> (Ibid.)

For Baker, this land is, as Spurr (1993, 31) would put it, 'already white, already home; colonists only have to accept what is theirs'.
  Baker continues:

> To carry out these ideas, it was necessary to set to work; and I determined to make a regular settlement at Newara Ellia, sanguinely looking forward to establishing a little English village around my own residence.
>   Accordingly I purchased an extensive tract of land from the Government at twenty shillings per acre. I engaged an excellent bailiff, who with his wife and daughter, with nine other emigrants, including a blacksmith, were to sail for my intended settlement in Ceylon.
>
> (Baker 1855, 14–15)

While Baker's settlement scheme failed, the town of Nuwara Eliya went on to flourish as a hill station and continued to enchant tourists and residents alike as a bit of Britain in the tropics.
  But utilitarian and romantic recuperations of the landscape could and did coexist in the same texts. Baker (1855, 32), now writing in the language of romanticism, for he is now viewing for pleasure rather than for profit, describes the view from an 8,000 foot peak near Nuwara Eliya thus: 'There is a feeling approaching the sublime when a solitary man thus stands upon the highest point of earth, before the dawn of day, and waits the first rising of the sun. Nothing above him but the dusty arch of heaven. Nothing on his level but empty space; all beneath, deep beneath his feet.' De Butts (1841, 161) also uses the sublime, but much more loosely when he writes, 'as one rises from the coast into the mountains, the country becomes at every step more romantic and wild. From the heights to the eastward of the town [of Kandy], the best views of this sublime landscape may be obtained.' However, being lower than the Himalayas, the Kandyan Highlands were more often described in the language of the picturesque than of the sublime.
  Writers commonly described the Highlands by comparing them to sites in Europe that were familiar to their readers. In doing so they engaged in an act of translation, recuperating the specificity of this place to a series of places on the tourist circuit in Europe. Baker (1855) described the lower parts of the Highlands as having 'an Italian climate' (11) while the higher altitudes were 'like a beautiful

cool mid-summer in England' (17). Matheson (1870, 175) informed readers that they could expect 'something like a combination of mountain views in Switzerland and Scotland'. The links to Switzerland, so popular with the British at this time, were frequent. Visitors to Kandy could even stay in the Hotel Suisse, which advertised itself in Cook's guide as having 'mountain scenery, a pictur-esque lake, and French food' (Cook 1923, 235). And yet the 'Orient', with its latent fears of disease, intruded on this fantasy of Switzerland on the equator in the form of a reference to the hotel being in 'the healthiest part of town'. Simi-larly, those who came for the air were enjoined to take early morning walks 'when the mountain air is keen and invigorating' (Cave 1912, 320), for by midday in Kandy there was no mistaking that one was not in Switzerland.

Most of the writers, however, compared the Highlands to Britain. One described the area between Kandy and Nuwara Eliya as having 'a glorious climate with dry invigorating air, like that of an English summer – away from the same-ness of the jungle and palm and heat of the plains' (*Ceylon Motorist's Road Book* 1916). Another described the area around Kandy as a 'Cumbrian panorama', and said that one could see from a hill above the town 'a glorified version . . . of our own Lake Country' (Farrar 1908, 63).

Yet another (Hurst 1891, 202) stated it more emphatically:

> In Kandy whether one will or not, the mind will go back to the Lake region in England. You find a calm and quiet beauty, freedom from strain and stress, a cluster of hillsides which throw down their beautiful face into the mirroring lake at their feet, a sweetness in all the pulsations of the air and a universal friendliness between all Nature and its lord, which brings up Grassmere, Windermere, Derwentwater, and their spirits – Southey, Wordsworth, Coleridge, and all the rest of the Cumberland immortals. Even the hostelry of Kandy, the Queen's Hotel, suggested to me immediately the Keswick Inn.

The language is interesting here, for it suggests that one had no choice but to see the place as a mirror of England (whether one will or not, as the author says). It asserts that the colonialist fantasy of the colonies as a part of Britain had come true in this place. And yet such a fantasy, the insistence that Kandy must be seen as the Lake District, could only work through a double erasure and inscription. First, a portion of the Kandyan cultural landscape was literally erased and inscribed with British buildings and plantings, and secondly those Kandyan features which remained (temples, palm trees, natives) were erased in the mind, with imaginary British landscape features and people inserted in their place. Such erasure and inscription, both on the ground and in the imagination, could only succeed when the visual and the aesthetic were privileged over other ways of understanding the place, which was precisely what the picturesque demanded. And yet the irony is that this privileging of the visual entailed the visual erasing in the mind of that which the eye could clearly see.

The native is continually 'overlooked', as Bhabha (1989) points out, in the double sense of being considered unworthy of attention and a danger to be constantly surveyed. Although one can read page after page of travel accounts with little or no mention of the local people who inhabit the country, occasionally the native is summoned to appear before the reader.[2] But the native is certainly not summoned in order to speak his mind (and it is nearly invariably he) but rather to be seen, to stand in for the Orient, which is to say to add to the picturesqueness of the scene. For example, at one end of the Kandy lake, this Windermere of the East, we are told 'the native delights to disport himself with water dashing over his dusky form' (Cave 1912, 320). Likewise, the local economy is aestheticized and transformed into a kind of playful, theatrical version of the British economy: 'The streets of Kandy will interest the visitor only insofar as they provide a glimpse of native town life and occupation in the bazaars; this is however always amusing to the visitor who is a stranger to Eastern customs' (Ibid., 323). Similarly that holy of holies, the Temple of the Relic of the Buddha in Kandy, is described as appearing 'in all its picturesqueness' (Woolf 1914, 76).

At times the shock of the familiar is accentuated by the juxtaposition of the native. Consider this account by an Englishwoman (Woolf 1925, 62):

> I saw in an upcountry garden an astounding sight – tufts of primroses growing round a tree – primroses just as wonderful as strew the banks and copses and fields in our English spring time. Then silently we stood and looked at them as if we could never look and gloat enough. Primroses! It is only when one is seven thousand miles away that one realizes what a thrill they can give you. How they stand for the essence of the wild, wet English spring with its gleam of pale gold sunshine. And memories of things seen come crowding on one – a wood in Sussex in the early days of February, with aconites and snowdrops pushing up in thousands through the dark earth – tufts of primroses found on Christmas Day in a Somerset copse, hidden in a mossy bank – white violets with rich warm scent blooming in a spinney on the Mendips, – meadow cowslips bordering the narrow stream in Cambridgeshire – and perhaps most entrancing of all, the lanes and meadows of Gloucestershire showing in their thousands and tens of thousands the nodding daffodils –
>
> 'Primrose vettu kondu va,' said our hostess to the garden coolie, and he set to work to pick the blossoms. The full incongruity of primroses in the tropics appeared at that moment. The primroses redolent of the English countryside, of cottage gardens, rosy-faced children, or flowersellers at street corners, thrusting 'penny bunches' into one's face, and the garden coolie with his gold nose-rings and ear-rings and scarlet handkerchief knotted round his head, his brown legs and tucked up cloth, shabby tweed coat, and the inevitable sack draped over his head. . . . One can feel a glow of Christmas home memories at the sight of holly and mistletoe. But it takes a primrose in the tropics to give one true nostalgia.

We can see here how an aestheticized vision operates. The primrose stands for a type of flower which in turn stands for spring in England. The chain of signifiers unwinds as we move from flowers to cool weather (the antithesis of the tropics with all of its negative connotations) to various English counties, to the countryside ideal, to cottage gardens, to English people – children and flower sellers. From the one signifier a whole world, 7,000 miles away, is conjured up. And yet, the reverie is shattered by the native. The incongruity of the primrose in the tropics is thrown into relief by the native. And yet it is unclear whether it is the native or the primrose that is incongruous. In truth, in this instance, both are British imports, for the 'native' is actually a Tamil, 'imported' by the British to work on the tea plantations that supported the British in the hills. But after noting the incongruity of the native and the primrose, the former is dismissed from memory and the nostalgic memory of England lingers on.

But while certain British things in the tropics conjured up surprise for the tourist and the joys of home memories for the British resident, other British things prompted outrage. By the 1880s, romanticism decreed that new buildings in Kandy were to be hybrid, that is they were to be British in size and form but were to have Kandyan ornamentation. Those that didn't came under heavy criticism. For example, the new government agent's office was criticized as 'an extensive and handsome building but alas! having no feature of any kind that harmonises with its surroundings. In an English manufacturing town it would not be out of place; but in Kandy it is a deplorable incongruity' (Cave 1912, 312). It is interesting to note that the English style which is deplored is not that of a pre-industrial countryside beloved of romantics but of a manufacturing city, thereby preserving the romantic image of British countryside alongside a similarly romantic image of the Orient. Even the Pavilion, the Governor's residence in Kandy, came under fire.

> It is fine in spaciousness but its furniture is uninteresting English stuff of the kind one finds in a Liverpool hotel, instead of the rich Oriental things they might so easily have had. . . . It made me sick to see the cheap Nottingham curtains in windows where Indian silks and embroideries naturally belong.
>
> (Rogers 1903, 252)

Once again we see a romanticism that unfavourably contrasts 'cheap' urban Britain to the 'rich' Orient. The argument here is that the British buildings and furnishings are inappropriate to the setting, that they aren't *'natural'*. But it doesn't seem to occur to these writers to extend this critique of the aesthetic to a broader claim that the British themselves might not belong in Kandy. Quite the contrary: their critique operates on the assumption that Kandy had been appropriated for their own visual pleasure and that they have not travelled so far to see cheap British goods. They have come to see the picturesque Orient and the Lake District of the east. This, then, is a critique based upon romanticism. It is utilitarianism that is being attacked here, things that are 'uninteresting' and belong in a 'manufacturing town'.

Another favourite image of travellers in search of the picturesque both in

Europe and Asia is social and physical decay. As one writer notes, 'Kandy makes as strong an appeal as any corner of the world. It is so calm, so gentle, so extinct. There is always here a ghostly afternoon, a perfect rest after anguish, a perfect forgetfulness of time and things' (Farrar 1908, 57). Similarly De Butts combines a picturesque with a utilitarian perspective when he writes,

> The plains [around Kandy] comprise a vast extent of beautifully undulating country, dotted here and there with groups of large and majestic trees, the intervals between which are open and entirely free from jungle. The whole bears a striking resemblance to an English park on a large scale, which would be complete but for the total absence of cultivation and of the dwellings of man. A deathlike stillness seems to reign over this apparently deserted valley, and contrasts strongly with the busy, animated aspect of waving corn fields and happy hamlets that adorn the smiling face of an English countryside.
>
> (De Butts 1841, 161)

Such romantic notions of the picturesqueness of Oriental decay were figured rather differently when expressed by the long-term colonial resident rather than by the tourist. For here the aesthetics of decay were reinforced by a belief in environmental determinism which suggested that Europeans who remained too long in the tropics would themselves decay and degenerate into hybridity – that is, become a mirror image of the Brown Englishman: the white native. Such thought was a nineteenth-century reworking of the environmental determinism propounded by Montesquieu in the eighteenth century. By the middle of the nineteenth century, race was seen as the key explanatory factor and environmental determinism was subsumed within it (Arnold 1996, 29–38). By the late nineteenth century, however, climatic concerns were once again highlighted within the discourse of race. Woodruff's theory of tropical neurasthenia, propounded at this time, was an attempt to explain why whites could not acclimatize to the tropics (Kennedy 1990; Livingstone 1992, 232–41). These ideas were repeated by American geographers such as Semple and Huntington in the early twentieth century. Huntington in *Civilisation and Climate* (1915) argued that in the tropics the physical, mental and moral constitution of the white race deteriorated. Such beliefs were not restricted to academics alone.

The rise and continued use of high-altitude hill stations in South Asia was based upon various strands of environmental theory from the early nineteenth century on (Kenny 1995). In such places, it was believed Europeans could somewhat escape the damage which the tropics caused to their race. Part of the enduring appeal of the Kandyan Highlands lay in the possibility which it offered of some such escape.

> To the enervated European residents of the plains of India it is a veritable paradise; they are discovering that a visit to Kandy and Nuwara Eliya is not only a source of health but of enjoyment, and that it restores their vanished

energies without the great expenditure of time and money involved in a voyage to Europe.

(Cave 1910, 1)

But this was, in fact, not the same as returning to Europe; it was a simulation. For the climate at Nuwara Eliya, and even more so at Kandy, only created an illusion of a European climate. As with the illusion of the English landscape in Ceylon, it needed the collaboration of the European. While the average tourist might be willing to succumb to this illusion, the long-term resident was decidedly ambivalent.

It is interesting in this regard to read the observations of a European visitor who refused to give in to the illusion. Ernst Haeckel was a German professor and devotee of Darwin who visited Ceylon in 1882. Like so many others he was at first struck by the similarity between the Highlands and northern Europe. He wrote, 'It is sometimes impossible not to fancy that one has been transported to the Scotch highlands, fifty degrees further north; and here in Newara Eliya [*sic*], precisely the same gloomy feeling came over me again and again as had possessed me when I traveled through that country in the autumn of 1879' (1882, 289). But, he argues, the similarity doesn't bear close inspection: 'The resemblance of this Promise Land of Newera Ellia [*sic*] to Northern Europe, though it has won for it the suffrages of European colonists, is, after all, but superficial for the most part, and closer observation reveals many differences. This is equally true of the climate and of the vegetation . . .' (Ibid., 293). He draws out the implications of such difference for health.

Figure 2 Painting of Nuwara Eliya, 1864, by Captain O'Brien.

The merits of Newera Ellia as a sanatarium are, in fact, monstrously over-rated. The climate is cold and damp; the temperature rises, on a clear winter morning, from about five degrees centigrade at dawn, to twenty-five or thirty degrees by noon. . . . This of course makes the visitor liable to sudden chills, and for rheumatic patients, or those susceptible to catarrh, it is perfectly intolerable.

(Haeckel 1882, 291)

Even in the Kandyan Highlands, the tropics have not been completely escaped. This point is made forcibly some years later by L.B. Clarence, judge of the Supreme Court of Ceylon, who spent twenty-seven years in that country. In a paper delivered to the Sixth Ordinary General Meeting of the Royal Colonial Institute in London he said,

The climate, though tropical, compares favorably with that of other tropical regions, and though our European race cannot thrive there in continuous residence for successive generations, an Englishman with due care may live there during a long and vigorous working life, especially if his livelihood lie in the hills or permit access to their cooler air.

(*Royal Colonial Institute Proceedings* 1895–6, 314)

I will leave the last word to a second-generation planter born in Kandy in the 1850s and writing his memoirs in the 1920s. He confirms Haeckel's and other people's view that although the Highlands may look like Europe, they are still in a fundamental sense the tropics. There appears no escape for whites who have lived too long in the tropics, even the high-altitude tropics.

Although I was born in an atmosphere of rigid British ideas; ideas that made me feel as thorough a Briton as if I had never been out of England, yet I lived in a climate that was tropical; surrounded with all the contracting influences, that unconsciously or subconsciously become Colonial. . . .

. . . and experience has often showed that the locally born was a far inferior creature; mostly dispised, generally distrusted, and invariably classed as 'country bottled', a particularly offensive appellation.

(Lewis 1926, 1, 32)

I have argued that the picturesque is a way of seeing that fosters hybridity, creates it, values it. The picturesque produced Anglo-Oriental landscapes in Britain and Asia. But this hybridity is figured differently for different types of British people. For tourists and those who created these landscapes in Britain and Ceylon the hybrid represented Britain extended. However, for second-generation white set-tlers, hybridity escaped the bounds of the aesthetic and spilled over into a related discourse of environmental determinism and cultural degradation. This move is figured as a shift from aesthetics to science, from the landscape to the colonial body, from exteriority to interiority, and from pleasure to anxiety. The shift from the picturesque to environmentalism, in other words, situates 'the third world within the first world' within the colonial body itself.

NOTES

1   This particular imaginative geography was the subject of my book, *The City as Text* (1990).
2   In fact there were very few Europeans in the Highlands. In 1910 Kandy had a population of 25,000, 100 of whom were European.

REFERENCES

Andrews, M. (1989) *The Search for the Picturesque: Landscape aesthetics and tourism in Britain, 1760–1800*, Stanford: Stanford University Press.

Arnold, D. (1996) *The Problem of Nature: Environment, culture and European expansion*, Oxford: Blackwell.

Baker, S.W. (1855) *Eight Years in Ceylon*, London: Longmans, Green and Co.

Bhabha, H. (1989) *The Location of Culture*, London: Routledge.

Campbell, J. (1843) *Excursions, Adventures and Field Sports in Ceylon*, London: Green and Co.

Cardinal, R. (1997) 'Romantic travel', in Roy Porter (ed.), *Rewriting the Self*, London: Routledge, pp. 135–55.

Cave, H.W. (1910) *The Ceylon Government Railway: A descriptive and illustrative guide*, London: Cassell and Co.

—— (1912) *The Book of Ceylon: An account of its varied attractions for the visitor and tourist*, Colombo and London: Cassell.

Ceylon Motorist's Road Book (1916) Colombo: Times of Ceylon.

Cook, T. (1923) *India, Burma and Ceylon: Information for travellers and residents*, London.

De Butts, L. (1841) *Rambles in Ceylon*, London: W.H. Allen.

Dougherty, J.A. (1890) *The East Indies Station, or the Cruise of H.M.S. Garnet 1887–90*, Malta: Muscat Printing Office.

Duncan, J. (1990) *The City as Text: The politics of landscape interpretation in the Kandyan Kingdom*, Cambridge: Cambridge University Press.

Farrar, R. (1908) *In Old Ceylon*, London: Edward Arnold.

Ferguson, J. (1903) *Ceylon in 1903*, Colombo: A.M. and J. Ferguson.

Gooneratne, M.Y. (1968) *English Literature in Ceylon 1815–1878*, Colombo: Lake House.

Gregory, W. (1894) *Sir William Gregory: An autobiography*, ed. Lady Gregory, London: John Murray.

Haeckel, E. (1882) *A Visit to Ceylon*, London: John Murray.

Head, R. (1986) *The Indian Style*, Chicago: University of Chicago Press.

Huntington, E. (1915) *Civilisation and Climate*, New Haven: Yale University Press.

Hurst, J.F. (1891) *Indika. The Country and the People of India and Ceylon*, New York: Harper and Brothers.

Kennedy, D. (1990) 'The perils of the mid day sun: climatic anxieties in the colonial tropics', in J.M. MacKenzie (ed.), *Imperialism and the Natural World*, Manchester: Manchester University Press, pp. 118–40.

Kenny, J.T. (1995) 'Climate, race, and imperial authority: the symbolic landscape of the British hill station in India', *Annals of the Association of American Geographers* 85, pp. 694–714.

Lewis, F. (1926) *Sixty Four Years in Ceylon*, Colombo: Colombo Apothecaries.

Livingstone, D.N. (1992) *The Geographical Tradition*, Oxford: Blackwell.

Matheson, J. (1870) *England to Delhi: A narrative of Indian travel*, London: Longmans, Green and Co.

Minh-Ha, T. (1989) *Woman, Native, Other*, Bloomington: Indiana University Press.

Rogers, K.R. (1903 [1976]) 'I am more and more delighted with this island and its people', in H.A.I. Goonetileke (ed.), *Images of Sri Lanka through American Eyes: Travellers in Ceylon in the 19th and 20th Centuries*, Colombo: International Communication Agency, United States Embassy, pp. 244–55.

Royal Colonial Institute Proceedings (1895–6) Sixth Ordinary General Meeting, vol. 27.

Said, E. (1979) *Orientalism*, New York: Vintage.

Spurr, D. (1993) *The Rhetoric of Empire: Colonial discourse in journalism, travel writing and imperial administration*, Durham, NC: Duke University Press.

Willis, J.C. (1907) *Ceylon: A handbook for the resident and the traveller*, Colombo: Colombo Apothecaries.

Woolf, B.S. (1914) *How to See Ceylon*, Colombo: Times of Ceylon.

—— (1925) *From Groves of Palms*, Cambridge: W. Heffers and Sons.

8 The Exoticism of the Familiar and the Familiarity of the Exotic

Fin-de-siècle travellers to Greece

Robert Shannan Peckham

> Italy is heroic, but Greece is godlike or devilish – I am not sure which, and in either case absolutely out of our suburban focus.
>
> (Forster 1908, 197)

INTRODUCTION: THE INSTABILITY OF CULTURAL DIFFERENCE

'Of all books', wrote Alphonse de Lamartine in his *Travels in the East* (1835), 'the most difficult, in my opinion, is a translation. Now, to travel is to translate' (Lamartine 1850, 82). Lamartine's reflection, inspired by the Parthenon's sublime 'chaos of marble', stands as a fitting epigraph for contemporary concerns that have focused on the relationship between translation and travel. In particular, attention has been paid across a broad range of disciplines to the processes through which unfamiliar cultures are translated in terms of the familiar; on the manner in which the exotic is deciphered and rendered intelligible by deploying recognizable conventions (see Boon 1982; Pratt 1992). The relations between travel and translation are further underlined by the etymology of 'translation', meaning 'carried from one place to another', which echoes the etymology of 'metaphor', a Greek word signifying 'that which is transported' (Hillis Miller 1995, 316; Butor 1974). If travel is a metaphoric practice, then it may be thought of as a form of writing, just as writing may reciprocally be conceived as a form of travel. As James Clifford has recently observed, if 'travel were untethered [and] seen as a complex and pervasive spectrum of human experience', then 'practices of displacement might emerge as *constitutive* of cultural meanings rather than as their simple transfer or extension' (1997, 3).

Travelling around the Kingdom of Greece some three decades after his compatriot Lamartine, Henri Belle, who served as First Secretary in the French Embassy between 1861 and 1863, visited Sparta, the capital of Laconia in the Peloponnese, which had been designed along neo-classical lines soon after independence from the Ottomans in the 1830s. The city had, the Frenchman noted, 'the aspect of a German spa town' (1881, 297). Commenting elsewhere on the rural inhabitants of mainland Greece, Belle remarked that in their hardiness they resembled Bedouin tribesmen, maintaining the same warlike customs. After all, Greece was, the traveller reflected, 'a little corner of the Orient', just as it constituted 'a little corner of Europe' (Ibid., 87).

Articulated in Belle's narrative were two oppositional models within which the independent Greek kingdom was mapped: in terms of stylized images of the West and of the East. In being evoked simultaneously, the authority of these categories was qualified. Greece was envisaged as a familiar, ordered European state, replete with European institutions, and ruled over since the overthrow of the Bavarian King Otto in 1862 by the Danish Holstein-Sonderburg-Glücksburg dynasty. At the same time, Greece, which had been an integral part of the Ottoman Empire for over four centuries and thus shared a common experience with much of the Arab Middle East, was conceived as an exotic, Oriental country of 'picturesque barbarity' (Gennadius 1892, vii). Writing a decade after Belle, in 1892, Gaston Deschamps asserted that 'Greece no longer occupies the lowest position among European nations. She is larger than Belgium and Holland together' (1892, 58). In the same breath, however, Deschamps could exclaim: 'Without sharing the pessimism of a traveller who called the Greeks "white Negroes", one can say that they [the Greeks] still have a lot of work to do before they resemble a civilised nation' (Ibid., 70–1).

If Greece could be mapped as a European kingdom, and defined in terms of progress, it could also assume a likeness with darkest Africa. Urging travellers to 'luxuriate in his last taste of European life' (Farrer 1882, 5) at the St George Hotel on Corfu, an island which had formed part of a British Protectorate of the Ionian Islands (1815–64) and bore evidence in 1880 of the 'tenacity of English habits and influence' (Ibid., 199), the Englishman Richard Ridley Farrer asserted that:

> Knowledge only comes with experience, and the one thing needful in contemplating such a tour [of Greece] is to abandon the fiction that the Hellenic kingdom has any pretensions to be called civilised. Let the pilgrim wait for warm weather, say in May or June, and make the same preparations as though he were about to go through Arabia or Central Africa.
>
> (Farrer 1882, 57)

According to George Warrington Steevens, a foreign correspondent for the *Daily Mail*, who was dispatched to India, Egypt and the Sudan, but who reported on the 'Thirty Day' Greco-Turkish war of 1897, the Greek kleftic fighters were 'just as the Sudanese and Zulus . . . more formidable with javelins than with rifles'. Had they relied upon their traditional fighting practices and not adopted foreign, Western tactics, he argued, the outcome of the war might have been successful for the Greeks (Steevens 1897, 285).

These mappings of Greece as Arabia and Africa reproduced an exotic topos that, at least by the late nineteenth century, had become as familiar as those projections of the European homeland against which the exotic was being pitted.

Conversely, the preoccupation in Europe with national cultures, which found expression in such practices as folklore (the Folklore Society in Britain was founded in 1874), increasingly shifted attention onto those unfamiliar elements within the national space itself. The 'vertigo and whirl', in Max Nordau's words, of modern, industrial society was degenerating the national character (Nordau 1895, 42; Pick 1996). Technologies linked with travel, such as railways, ships, telephones and newspapers – technologies which supported the running of empire – were also bringing about the downfall of 'civilized life'. In the 1890s, Nordau declared, 'a petty tradesman travels more and sees more countries and people than did the reigning prince of other times' (1895, 39). Travel was both a cause and a conspicuous symptom of social instability, while the city, as the mainspring of such degenerate tendencies, was pitted against sedentary rural communities. It was only in the wild, outlying, rural regions of the nation-state, far from the onslaught of cosmopolitan modernity, that authentic national traditions lived on.

If exotic places were being translated in terms of the familiar, familiar places were being correspondingly mapped in terms of the exotic (see de Certeau 1986, 119–67). In France overseas colonies provided an analogy for the undomesticated provinces and furnished a comparative framework for the description of rural French customs and idiosyncrasies (Weber, 1979, 485–96). The exotic was not only an extraneous, *colonial other*, against which European authority was defined; it was also an internal category that marked out healthy differences within what Richard Claverhouse Jebb, Professor of Classics at Glasgow, and later Regius Professor of Greek at Cambridge, called, in a lecture on Greece delivered before the Philosophical Institution at Edinburgh in 1880, an 'organically composite nationality' (Jebb 1880, 50).[1] Similarly, in his popular travel book *Rambles and Studies in Greece* (1892), the very title of which is reminiscent of British folkloric preoccupations, John Pentland Mahaffy suggested that a voyage to Greece was a means of reconnecting with a genuine national character which was in danger of disappearing before a 'great tide of sameness':

> If he [the traveller] desires to study national character, and peculiar manners and customs, he will find in the hardy mountaineers of Greece one of the most unreformed societies, hardly yet affected by the great tide of sameness which is invading all Europe in dress, fabrics, and usages.
>
> (Mahaffy 1892, viii)

Much the same argument was put forward by the explorer and archaeologist James Theodore Bent, who studied the local traditions and customs on the Aegean islands in the late 1880s. In a description of a visit to Karpathos, he observed that 'no happier hunting ground could exist for the study of unadulterated Greek peasant life' (Bent 1886, 199).[2] According to the Baedeker *Handbook*, Greece was a haven for those 'wearied of the artificial and overcivilised side of modern hotels and means of locomotion' (Baedeker 1889, xxiii; see also Buzard 1993).[3]

EUROPEAN, ORIENTAL, OR BALKAN?

Was Greece European, Oriental, or Balkan? Did it belong to 'the Mediterranean', a geographical unity promoted with the publication of Murray's *Handbook to the Mediterranean* (Playfair 1881), large sections of which were devoted to Greece?

The preoccupation with defining Greece as a geographical and cultural space needs to be considered within contemporary debates about the shifting meanings of 'Hellenism' (Turner 1981), as well as within debates about degeneration and the imperative to demarcate the bounds of national cultures. In Britain concerns about the corruption and degeneration of national life coexisted and were not unrelated to drives to open up and define new territories. The idea of the nation as a 'limited sphere' (Anderson 1991, 7) was inextricably bound up with the colonial Other against which the modernizing nation was defined. In 1907 Lord Curzon of Kedleston, in his Romanes Lecture at Oxford, alluded to Turner's frontier thesis and invited his audience 'to pause and consider what Frontiers mean, and what part they play in the life of nations' (Curzon 1907, 5), calling for studies to be made of 'Border literature' (Ibid., 54). Curzon had noted that 'every Greek war is waged for the recovery of a national Frontier' (Ibid., 6), while Belle himself had likened Greece to a European frontier territory – a more familiar equivalent, as he put it, of the American Far West (Belle 1881, 105). In fact, Greece constituted an important subject for such 'border literature' since it was conceived both as the source of Hellenism and as a vital geopolitical space in the establishment of a European bulwark against the encroaching East. As Edward Said has noted, Orientalists feared 'not the destruction of Western civilization but rather the destruction of the barriers that kept East and West from each other' (1978, 263). Lewis Sergeant, the author of *New Greece* (1878) and the editor of the *Educational Times*, remarked in 1897, 'If there is a single Englishman who is not convinced of the real importance of Greece to Europe . . . he would do well to sit down before a map of Europe, to blot out the name of Greece, and to substitute for it "Turkey", "Slavonia", or one of the letters of the "Russian Empire"' (Sergeant 1897, 2).

Greece therefore occupied an important position in the debate about the relations of the West to the East. In 1901 Meredith White Townsend, editor of *The Spectator*, published a collection of articles and reviews entitled *Asia and Europe* in which he strove to demonstrate that the East is separated from the West 'by a chasm that nothing can bridge' (Bevan 1903, 63). Townsend's book provoked a number of counter-invectives, among them an article by Edwyn Robert Bevan, the author of *The House of Seleucus* (1902), who lectured on Hellenistic history and literature at King's College London – and whose brother, Anthony Bevan, taught Arabic at Cambridge. In it he asserted:

> *The East!* The popular writer knows well that it is a term to conjure with. One can hardly open a magazine without finding some allusion to 'the East', 'the immemorial East', 'the brooding East', 'the mysterious East', 'the unchangeable East'. These phrases never fail to awake the appropriate thrill, the rush of

associations. And this inscrutable entity, the popular theory goes on to affirm, is separated from 'the West' by a chasm that nothing can bridge. . . . The genesis of this current view it would be interesting to see traced. It is in part due, no doubt, to a *literary* tendency towards a telling and picturesque way of representing things: the poets have done their parts.

(Bevan 1903, 62–3)

In effect, the purpose of Bevan's critique was to assault Townsend's position by exposing 'the widely accepted belief in certain essential and innate difference of character, temperament and moral which separate the Oriental' from the British (Clifford 1903, 128).[4] Placing inverted commas around such terms as the Asiatic, the Oriental and the West, Bevan aimed to show how such notions were abstractions grounded on nothing more substantial than the negation of European qualities (Bevan 1903, 67). His emphasis on the power of representations and his call to 'trace the genesis' of those potent stereotypes that had accumulated around an idea of 'the East' resonate poignantly within contemporary critiques of Orientalism.

Such debates about the differences and congruities between the Orient and Europe, about the location and legitimacy of those borders that marked off distinct cultural and geographical enclaves, provided an important context within which travellers endeavoured to define Greece. They further underline the extent to which travelogues were shaping, and reciprocally being shaped by, broader debates conducted in newspapers, periodicals, as well as political and cultural surveys. Travel writing conceived as a general description of a foreign place for home consumption was giving way to more specific kinds of writing: to journalism, geographical, political historical, cultural and economic treatises. In this way, the travel guide functioned as an envelope for a diverse and complex ensemble of texts. R.C. Jebb divided his Edinburgh lectures into a discursive analysis of Greek national character and history, and 'the impressions of a visit to the country, so far as these may serve to illustrate the chief traits of its present aspect and condition' (Jebb 1880, 16). Often the framework within which such categories as 'Asia' or 'Europe' were deployed was consciously articulated by travellers. Agnes Smith, one of the first female travellers to have visited the Morea, in her *Glimpses of Greek Life and Scenery* (1884), for example, declared that 'Opinions must of course differ according as we judge them [the Greeks] by an European or an Asiatic standard' (341).

In many travelogues the Greeks were conceived as dwelling 'in a twilight zone illuminated neither by the radiance of the West nor by the exotic glow of the East' (Augustinos 1994, 285). Jebb himself had sought to distinguish between two types of Greeks: the hybrid Greeks of Asia, and the Greeks of Europe. At the same time, the shifting stereotypes of the familiar and the exotic employed to describe Greece troubled the security of those same stereotypes. The dis-orientation evident in travelogues on Greece, it could be argued, sheds useful light on the dis-orientation of twentieth-century ethnography, in which, as Clifford has observed, ' "cultural" difference is no longer a stable, exotic otherness', and where a 'structure of expectations about authenticity in culture and in art is thrown into doubt' (1988, 14).[5] Greece stood for shifting notions of cultural difference: difference within a type, and differences between imputed types (Todorova 1997, 19). In the first case, Greece was construed as belonging along with the rest of the Balkans to what Allen Upward called 'the East End of Europe' (1908).[6] In the second case, Greece was conceived as belonging to the East as 'a little corner of the Orient'.

CLASSICAL AND MODERN GREECE

Greece's authenticity as a sovereign state rested, in great measure, upon its assumed inheritance of ancient Greece. If Hellas was, in the words of William Makepeace Thackeray, 'the most classical country in the world' (1844, 52), it was a legacy claimed by all European nations, including the British, French and Germans, who had established archaeological schools in Athens with the specific aim of uncovering the glories of the classical past.[7] 'In all parts of the country', Murray's *Handbook* declared, 'the [foreign] traveller is, as it were, left alone with antiquity' (1854, 21). 'Greece has', the fourth edition of the *Handbook* remarked, 'no modern history of such a character as to obscure the vividness of her classical features' (1872, 2). As one British author observed, adding his voice to the pleas of numerous other travellers who called for a British school to be established in Athens, 'If any justification were necessary, they [the Europeans] might even claim to be in a sense themselves heirs to the ancient civilisation of that country, so completely has it impressed itself on the minds and habits of thought in England and France' (*Quarterly Review*, Jan/Apr 1885, 298–322). In a notice that referred to Lord Bute's generous donation of £500 towards the creation of a British School,[8] *The Times* asserted that 'a general feeling exists that England ought not to remain behind France, Germany, and the United States in the advantage of possessing a permanent centre in Greece'.[9] The purpose of such an institution would be 'to afford information and advice to properly accredited British travellers in Greece', while 'through the agency of the School, valuable notes might be collected from visitors to the Hellenic countries, who, without being specialists, are competent scholars and observers'. One such traveller who benefited from the agency of the School was the anthropologist James Frazer, who travelled to Greece in 1890 and 1895 to prepare an edition of Pausanias' *Description*.

The extent to which concerns about the Eastern or Western origins of ancient Greek culture were bound up with contemporary preoccupations about the political and geographical limits of Europe and Asia is manifest in William Mitchell Ramsay's survey, *The Historical Geography of Asia Minor* (1890). Ramsay was a classical scholar, archaeologist and expert on the topography and history of Asia Minor in ancient times. In the opening section of his study, entitled 'Hellenism and Orientalism', the author maintained that throughout history Asia Minor has been 'a battlefield between the East and the West'.[10] At the same time, however,

Ramsay argued that if 'the religion, art, and civilisation of the East found their way into Greece', Hellenism passed back into the East with Alexander the Great's conquests in the fourth century BC. 'The very character of the country', he noted, 'has marked it out as a battleground between the Oriental and the European spirit' (Ramsay 1890, 23).

By turns, Ramsay undermined and affirmed ideas of an essential difference which underpinned East and West, European and Oriental. The late Byzantine period, he asserted, was characterized by its 'Oriental character' and each of the imperial dynasties 'was less "Western" than the preceding one' until 'at length a purely Oriental dynasty of Osmanlis eliminated even the superficial forms of the West'. As if acknowledging the contingency of the category he is mobilizing, the adjective 'Western' is placed within inverted commas in the text. Taking stock of the contemporary situation, Ramsay declared that 'the Greek element is gradually supplanting the Oriental on the Aegean coast', even though, he added, 'the interior is still wholly Oriental' (Ibid., 25). In Ramsay's narrative, therefore, Greece is both a product of the East and a buffer to its decaying influence. Hellenic culture is envisaged simultaneously as being composite, formed by reciprocal borrowings, and as a pure, Occidental transhistorical essence, doing battle with the Orient in perpetuity.

Beyond the specific concern with Hellenism, Ramsay's analysis raises larger questions about the defining characteristics and limits of a culture, just as the shifting attitudes of travellers to Greece expose a set of interrelated assumptions, not only about the place of modern Greece, but also about the location and bounds of Europe. This equivocation is reflected in the structure of the guide-books themselves, which sought to establish a difference between the traveller and the populations of the country he/she visited, and yet at the same time strove to promote an identification with the 'native' populations among whom he/she was travelling. As Baedeker's *Handbook* declared, on the one hand the purpose of the guide was to protect the 'unsuspicious' traveller from falling 'prey' to the wiles of the locals (Baedeker 1889, liii). On the other hand, as the original 1840 edition of Murray's *Hand-Book for Travellers*, the first comprehensive guide to Greece, compiled by the diplomat Henry Headley Parish, asserted, citing a quotation from the David Urquhart's *Spirit of the East* (1838):

> when you put yourself in a position similar to theirs, you can feel as they do, and that is the final result of useful investigation. Burke, in his essay on the 'Sublime and Beautiful', mentions an ancient philosopher who, when he wished to understand the character of a man, used to imitate him in every thing, endeavoured to catch the tone of his voice, and even tried to look like him: never was a better rule laid down for a traveller.
>
> (*Hand-Book* 1840, ii)

In short, the guidebook protected the traveller, distancing him from the 'strange peoples' he moved among, even while it sought to instruct the traveller into foreign ways, and to place him 'in a position similar to theirs'.

TRAVELLING THEORIES

It is perhaps because of the shifting conceptions of cultural otherness in the texts on Greece produced by European travellers that Greece is so conspicuously absent from Edward Said's study of Orientalism (1978). If, as Said maintains, Orientalism is a pre-eminently textual affair, many of the travel narratives on which he focuses, such as those by Flaubert, Chateaubriand, Nerval and Lamartine, devote important sections to Greece, sections which are noticeably excised from Said's account. Greece is ignored because it fails to fit in with a history that is defined solely in terms of colonial occupation of the Orient by the West, just as German Orientalism is dismissed because, unlike British and French Orientalisms, it was not so directly enmeshed in colonial domination. When Greece appears in *Orientalism* it is as fifth-century Athens, with Aeschylus' play *The Persians* and Euripides' *The Bacchae*, works which, Said claims, set off the Orient from the West and contain 'the essential motifs of European imaginative geography' (Said 1978, 56–7). Ironically, in reclaiming classical Greece as part of an essential and transhistorical category interchangeably called Europe and the West, Said mirrors the rigidity of the framework which he is ostensibly critiquing, and comes close to replicating the contradictions in Ramsay's *Historical Geography*.

Modern Greece is jettisoned from the argument precisely because it disturbs the notions of continuity upon which Said's diachronic reading of a European Greece rests. Greece unsettles the binary oppositions promoted in the book and exposes the inconsistency of Orientalism as a discourse that is characterized by its systematic nature and by a 'knitted-together strength' (Said 1978, 6). Christian Greece, however, as an indeterminate *space between*, calls into question many of the assumptions behind Orientalism,[11] and disturbs Said's attempt to balance a genealogical conception of intellectual history, indebted to humanist scholars such as Erich Auerbach, with a Foucauldian concern for discourse and the discursive field (see Clifford 1988, 255–76).

With a few exceptions, the orientation of post-colonial studies has tended to be exclusively towards the experiences of the Asian subcontinent and of selective North African states, in particular Egypt. What, however, of Ottoman imperialism? What of other forms of imperialism that cannot be associated exclusively with the European nation-state in the nineteenth century? The territory which became Greece at the beginning of the nineteenth century had constituted part of the Ottoman Empire for some four hundred years – a common imperial legacy shared with much of the Arab Middle East. For much of the nineteenth and twentieth centuries, Greek history can be read as a struggle for the Greeks to take their place, forcibly, as historical subjects.[12] Can Greece, therefore, be reclaimed for post-colonial studies?

Greece, it might be argued, offers an example of a two-fold colonization by the Islamic Ottoman Empire and by Europe. George Cochrane concluded his two-volume *Wanderings in Greece* (1837) with a chapter entitled 'The colonization of Greece', in which he detailed the financial opportunities available for entrepreneurial Englishmen in 'the waste lands of Greece', by employing 'natives of the soil' as labourers (quoted in Bastea 1997, 61). In fact, from 1815 to 1864 (when they were handed over to Greece) the Ionian Islands had constituted part of a British Protectorate, while the largely Greek-populated island of Cyprus had fallen under British rule in 1878, following the Congress of Berlin. After Greece's independence Otto of Wittelsbach, the son of King Ludwig I of Bavaria, was chosen by the so-called 'Great Powers' and installed as absolute monarch in a state which was modelled along European lines and 'guaranteed' by these same Powers. If constructions of the Greek past were imported into Greece from without, Greek writers sought to define a national culture from within these institutionalized European discourses of the past, but within structures inherited from centuries of Ottoman rule.

Many of the issues raised in a history of late nineteenth-century Greece have implications for the scope of post-colonial studies, which have largely ignored these marginal areas of familiar exoticism and have played down the heterogeneity of the colonial experience because it complicates the picture of Orientalism as a systematic discourse (see Ahmad 1992, 157–219).[13] Travelling around Greece Europeans envisaged themselves remaining within the bounds of a familiar culture *at the same time* as Greece was defined in dichotomous terms as an exotic, Asiatic Other.

READINGS

In his account of nineteenth-century Greek irredentism, an account which he contrasted to 'picturesque' travel records (Sergeant 1897, 2), Sergeant argued that the history of Greek expansion depended 'on no mere tradition, no laborious assemblage of texts and obscure references' (Ibid., 56). For travellers in the nineteenth century, a voyage to Greece was mediated through a canon of ancient texts which shaped whatever was written about the country. A glance at any guidebook or travel account of the period, with its copious exegetical passages devoted to Homer, Aeschylus and Pausanias, demonstrates the extent to which travel was inseparable from textual, interpretative processes.

Travel was prompted by readings, and the choice of places visited was made largely through reference to classical texts, just as these same texts determined archaeological sites. Schliemann confided, for example, that he had been inspired by his readings of Homer to dig at Ithaca, Mycenae and Troy. At least since the late eighteenth century, travel to Greece had been organized by, and given meaning through, readings of specific texts – and, in particular, Homer.[14] Topography authenticated the texts, so that Murray's *Handbook* declared, 'Greek authors acquire new and clearer meanings read by the light of Greek scenery and topography' (*Handbook* 1896, xxix). The landscape itself was envisaged as an artefact and defined in terms of its legibility, while the traveller was cast in the role of reader who activated the meanings of the manuscript: 'The aspect of Greece is that of the old manuscript: covered as it may be by many a palimpsest but it is only in proportion as the original text is read that the value is felt' (*Handbook* 1884, 9).

'No one can pretend to understand Greek history and the particular influences that contributed to mould the genius of its people', observed Murray's *Handbook*, 'without a clear comprehension of the hardy features of its geography' (*Handbook* 1896, lxvii). Yet if the 'scenery and topography' heightened the value of the ancient texts, it also drew attention to the gulf between the secure meanings of Greece as the origin of European culture inscribed in 'the old manuscript' and the inhabitants of the fledgling Greek kingdom. 'To travel to Greece', remarked one commentator in the *Contemporary Review* (1889), 'is to make two journeys; the one in the present, the other in the past' (D'Estournelles 1889, 586).[15]

At no time were the discontinuities between Greek places and people and 'the superficiality of civilization' in Greece (Wyse 1871, 18) more brutally exposed than in the spring of 1870, when a party of English aristocrats, including Frederick Vyner, the brother-in-law of the Lord President, Earl De Grey, and Edward Herbert, a cousin of Lord Carnarvon, the Tory statesman, were abducted by brigands on a trip to the ancient site of Marathon some 40 kilometres north-east of Athens. In the event four of the hostages were murdered and the ensuing furore in the British Press exposed the European stereotype of the Noble Greek, focusing on the discrepancy between ancient Greek heroism associated with the battle of Marathon, where the Greeks fought the invading Persians, and the perfidious behaviour of the brigand band. As *The Times* opined, Greece was a disgrace to civilization, a country of half-Greek demi-savages, which had to be punished, either by being occupied by Indian troops, by being handed back to Turkey, or even, in the last resort, by being abolished altogether (Jenkins 1961, 79).[16] The murder of the British tourists exposed the chasm between the noble past and the debased present, drawing attention to the 'contrasts of magnificence and meanness, of loftiness and lowness'. Ironically, the site of the 1870 ambush became a tourist site pointed out and noted by subsequent visitors, while the furore prompted a resurgent interest in Greece. Yet as Deschamps declared in 1892, the modern Greek continued to be considered by travellers as 'a jarring accident, thrown out of place among the sacred ruins of ancient Greece to spoil the spectacle and the impression'. In the 'colourful and motley Orient' which foreigners had invented, the Greeks appeared 'a little too magnificent and the Turks a little too Tartar' (quoted in Bastea 1997, 56–7).

ENGLISHMEN, GREEKS, TURKS AND ROMANS

Guidebooks and travel accounts of voyages to Greece focused repeatedly on the relative differences and similarities between Greeks and Europeans, or Greeks and Turks. If the Greeks were portrayed as imperfect Orientals, they were also imperfect Europeans and even in Athens where, as A. Proust noted, 'the Western way of life is deeper', nevertheless, 'it [was] a matter of surface treatment, not substance' (quoted in Bastea 1997, 59–60).[17] As Murray's *Handbook* commented:

> The habits and customs of the Greek peasantry may, in many instances, be traced back into classical times. That their manners are almost identical with the Turks, except in those points in which their respective religions have given rise to a difference, may be attributed to the strong tincture of Oriental customs which is traceable in the Greeks of every age, a consequence of their situation on the borders of the Eastern World.
>
> (*Handbook* 1896, 63)

The niece of Sir Thomas Wyse, British ambassador in Athens from 1849 to 1862, in the introduction to his *Impressions of Greece*, declared: 'Servility, an inordinate love of riches, and a want of truth – vices which engender the absence of honour and rectitude – are the chief sins justly laid to their [the Greeks'] charge' (Wyse 1871, 15). Because of the country's 'situation on the borders of the Eastern World', Greeks resembled but could be distinguished from the Turks, just as they could be defined through their proximity to Europe. Agnes Smith noted that 'betwixt her [Greece] and Turkey the difference is immense. The Greeks may have great faults, but they almost alone of Eastern nations are perfectly accessible to the influence of Western ideas' (Smith 1884, 341).

On the one hand, therefore, the Greeks were different from the Turks, since, as Murray's *Handbook* expressed it, 'no pressure of foreign domination, no admixture of alien blood, has sufficed to obliterate the old fundamental lines – for good and for evil – of the Greek character' (*Handbook* 1896, xxix). Whatever 'faults of character' were discernible in the Greeks were owing 'to the long servitude of the nation under Turkish rule' (Ibid., cvi). Baedeker's *Handbook*, however, entreated travellers to keep back payment from their interpreters 'as a spur to the inborn Oriental indolence of the Greek' (1889, xiv). The country was both characterized by its accessibility and simultaneously lauded for its hardiness.[18] As Murray's *Handbook* declared, 'As a general rule, the traveller should bear in mind that the unavoidable discomfort of travelling in Greece is so great, that it is desirable to have as few unnecessary sources of it as possible.' Greece was of interest for travellers but not a place 'for a mere idler or man of *pleasure*' (*Handbook* 1872, 1). Moreover Z. Duckett Ferriman remarked of the Greeks at the beginning of the century:

> The Greek is racially and geographically European, but he is not a Western [*sic*]. That is what he means by the term, and the signification is accepted by both Greek and foreigner. He is Oriental in a hundred ways, but his Orientalism is not Asiatic. He is the bridge between the East and West . . .
>
> (quoted in Todorova 1997, 16)

Greeks – at least the ancient Greeks – were not, in fact, that dissimilar to the English: 'The fact is that the Greeks, like the modern English, carried war in one hand and industry in the other' (Sergeant 1897, 57). The ancient Greeks had taken 'the one great step from the stationary into the progressive form of society; the advance from the darkness of Asiatic barbarism into the light of

European civilization' (*Handbook* 1872, 42). In his *Lectures on the Geography of Greece* (1873) Henry Fanshawe Tozer similarly asserted that 'Greece occupied in ancient times a position in many respects similar to that of England at the present day' (Tozer 1873, 5). This was an analogy that pervaded accounts of visits to Greece as well as surveys such as Sergeant's *Greece*, which appeared in a series of books on 'Foreign Countries and British Colonies', where the author asserts that the Greeks' 'claim to the title of Hellenes is scarcely in any respect inferior to our own title to the name of Englishmen' (Sergeant 1880, 73). Similarly, Greece was penetrated by 'Slavonian invaders', just as the Britons were 'degraded' under the Saxons and the Saxons under the Normans (Ibid., 77). More practi-cally, Baedeker recommended that the prospective traveller to Greece should pro-vide himself with a suit of grey tweed, such as is used by sportsmen at home' (1889, xvii), while Murray's *Handbook* cautioned, however, 'It would be ridicu-lous in any English traveller to assume the Greek or any other Oriental dress . . . he will still find an *English shooting-jacket and wide-awake* the most respectable and respected travelling costume throughout the Levant' (*Handbook* 1872, 6).

The identification of Englishmen with ancient Greeks was made by Virginia Stephen (later Woolf) in a short sketch inspired by a visit to Greece in the autumn of 1906 with her sister and brothers. In the sketch Woolf focuses on a party of English tourists on a trip to Mount Pentelicus, where the marble quarries for the Parthenon were located:

> It so happened not many weeks ago that a party of English tourists was descending the slopes of Mount Pentelicus. Now they would have been the first to correct that sentence and to point out how much inaccuracy and indeed injustice was contained in such a description of themselves. For to call a man a tourist when you meet him abroad is to define not only his circum-stance but his soul, and their souls they would have said – but the donkeys stumble so on the stones – were subject to no such limitation. Germans are tourists and Frenchmen are tourists but Englishmen are Greeks.
>
> (Woolf 1906, 979)

Woolf's allusion to the English Greeks recalled Matthew Arnold's notions of the similarities between the 'culminating' ages of Periclean Greece and Victorian England (Super 1960–77, 22). Significantly, in 1905 a heated debate was going on in Britain about the compulsory teaching of Greek in schools and the obliga-tory requirement of Greek for entry to Cambridge.[19] For Virginia Woolf the achievements of ancient Greece were despoiled by the modern inhabitants, as she noted in her Greek diary:

> You must look upon Modern Greek as the impure dialect of a nation of peasants, just as you must look upon the modern Greeks as a nation of mongrel elements and a rustic dialect of barbarous use beside the classic speech of pure bred races.
>
> (quoted in Lee 1996, 229)

While the analogy between imperial Britain and Greece was pervasive, the British were also frequently compared with Romans. Thus, Evelyn Baring, Lord Cromer, in his *Ancient and Modern Imperialism* (1910), exclaimed that 'the imperial idea [was] foreign to the Greek mind' (quoted in Jenkyns 1980, 333). The analogy was explicitly stated by Lord Palmerston, the British foreign secretary, in the Don Pacifico incident of 1847, when anti-Jewish riots in Athens led to the plundering of the property of Don Pacifico, a Maltese Jew who was Portuguese vice-consul and a British subject. When compensation was not forthcoming, Palmerston ordered a blockade of Piraeus in 1850, quoting the words 'Civis Romanus sum' (Clogg 1986, 79). Similar gun-boat diplomacy was enacted in 1885, in the notorious Nicolson affair. Nicolson, the Chargé d'Affaires at Athens, was out walking on the slopes of Mount Likavitos when he 'was wantonly assaulted by a gendarme, who struck him with a stick'.[20] Britain exacted heavy reparations, which included the gendarme's expulsion from the service, the British national anthem being played in Constitution Square with the Union Jack raised and the entire Athenian police force presenting arms.[21] As such examples show, British claims to be descendants of ancient Greeks were tempered by a belief in Roman power. Moreover, the view of the British as both Greek and Roman was translated into an idea of a British empire which possessed what Charles Adderly called, in 1869, its distinct 'Grecian' and 'Roman' elements (Jenkyns 1980, 333–4).[22]

A STRANGE LIKENESS: GREECE AND IRELAND

The manner in which Greece was mapped as an indeterminate space was not dissimilar to Ireland's mapping as 'a flawed version of England' (Kiberd 1996, 14; Kiberd 1997). Townsend argued in *Asia and Europe* (1901) that the Irish were in fact Asiatic, bearing a striking resemblance to the Egyptians: 'The Egyptians are not a strong people, but it is quite useless to tell an Egyptian that the Europeans bring him prosperity and light taxes, as useless as to tell a true Irish nationalist the same thing about the English' (Townsend 1901, 376). Arriving in Greece on board the *Iberia* Thackeray had been struck by the grimness of Athens, and in particular of the royal palace, commenting that 'the shabbiness of the place actually beats Ireland, and that is a strong word' (1844, 50). He asserts that 'one is obliged to come back to the old disagreeable comparison of Ireland and Athens', with streets swarming with 'idle crowds', 'dirty children', and women of 'sallow, greasy, coarse complexion, at which it [is] not advisable to look too closely' (Ibid.). This was, in fact, an analogy developed by contemporaries. Thus, Farrer concluded his *Tour of Greece* with an appendix containing 'useful information to intending travellers'. Among other advice proffered, he urged travellers:

> always to have a revolver ready to hand. Of course weapons are useless on encounter with brigands, who, like Irishmen, never attack except from a position of vantage, and with a numerical superiority over their victims of at

least ten to one; but the sight of an English-made pistol will overawe the cupidity of many an amateur robber, and act as a wholesome check upon the natural insolence of the population.

(Farrer 1882, 216)

Similarly, Jebb noted that the Greek horse was strikingly similarly to 'his Irish brother [which] has the knack of somehow getting over stone walls' (1880, 16). E.M. Lynch made the connection in 1900 in the pages of the *Gentleman's Magazine*, in an article entitled 'Greece and Ireland'. Beginning with an anecdote about an acquaintance who has just returned 'from a month's wanderings in Greece', Lynch endeavoured to trace the 'manifold points of likeness' between Greece and Ireland, which included traces of ancient civilizations, the importance of the church, patriotism, the love of litigation and folklore.[23] Finally, like Ireland, nineteenth-century Greece saw a massive wave of emigration, chiefly to the United States, an exodus which carried with it the idea of a national homeland and provided a vital economic boost to the country in the form of remittances (Lynch 1900).

A decade before Lynch's article, at around the time when Douglas Hyde and the Gaelic League (1893) had sharpened the focus of the debate about Irish national identity, another British commentator exclaimed, 'We might as well set Ireland adrift by itself in the middle of the Atlantic, and expect it to rival the British Empire, as expect the tiny Greek kingdom, with its two million inhabitants, to take her position as a power in the East'. Greek governments rise and fall, the same writer continued, 'with a rapidity startling even in a South American republic'. Fortunately, however, the reforming prime minister Charilaos Trikoupis was educated at Harrow; his father was Greek Ambassador, 'and made himself thoroughly master of our system of legislature, and under his guidance it is being brought to a successful issue on the old soil of Hellas' (Bent 1891, 290). Published at the height of the Home Rule debate, the analogy of Greece with Ireland is pertinent and recurs in travellers' accounts that depict the Greek landscape as an Irish wilderness of glens and barren mountains.

In his article, Lynch had referred with approval to Mahaffy as an authority on Greece. Mahaffy, the author of *Rambles and Studies in Greece*, which was reprinted in four editions between 1876 and 1892, taught Greek at Trinity College, Dublin, where he was Oscar Wilde's tutor, and contributed, along with Wihelm Dörpfeld, and Agnes Smith, to Baedeker's *Handbook*. At the beginning of his guide, Mahaffy declares that 'there is no country which can compare with Greece' (1892). It was a sentiment that ran through all the guidebooks. Thus, Murray's *Handbook* declared: 'the very scenery of Greece has a national character of its own', although the flora and fauna were consistently characterized through their affinity with European countries and Asia (*Handbook* 1896, lxiv). Throughout Mahaffy's book assertions of Greece's uniqueness are undermined by comparisons with familiar landscapes. The use of the noun 'rambles' in the title is reminiscent of the rediscovery of popular, rural customs and traditions in Britain where familiar, local cultures were increasingly being written about with the interest hitherto afforded to exotic cultures. The traveller, Mahaffy observes, 'will find in the Southern Alps and fjords of Greece, a variety and richness of colour which no other part of Europe affords'. Here, even while 'Greece' is represented as unique, the vocabulary employed ('Alps and fjords') casts the country as a southern version of northern Europe (Mahaffy 1892, viii).

In particular, a correspondence between Greece and Ireland runs through Mahaffy's rambles. 'I can never forget', he reflected, 'the strong and peculiar impression of that first sight of Greece; nor can I cease to wonder at the strange likeness which rose in my mind, and which made me think of the bays and rocky coasts of the west and south-west of Ireland' (Ibid., 4). The Acropolis rising above Athens is compared to an Irish landmark, the Rock of Cashel (Ibid., 105–6), while the ruins at Mycenae are likened to 'the great sepulchral monuments in the country of Meath' (Ibid., 415), and the illuminated manuscripts held in the monastic libraries on Mount Athos resemble the eighth-century *Book of Kells*. 'I have always thought it likely', Mahaffy remarked, 'that some early Byzantine missionary found his way to Ireland, and gave the first impulse to a local school of art' (Ibid., 465–6).

During the spring of 1877, on a visit to Tripolitsa in the Peloponnese, in the company of Wilde and George Macmillan, the son of the publisher, Mahaffy had been feared captured by brigands (Ellman 1987, 70). Yet in the 1880s and 1890s brigandage was brought under control under the premiership of the old Harrovian, Trikoupis. As Mahaffy noted in his *Rambles:*

> And yet, in spite of the folly still talked in England about brigands, he [the traveller] will find that without troops, or police, or patrols, or any of those melancholy safeguards which are now so obtrusive in England and Ireland, life and property are as secure as they were in our most civilised homes.
>
> (Mahaffy 1892, 426)

For Mahaffy Greece had inspired a feeling of 'strange likeness' with Ireland. While it resembled Ireland, it was also fundamentally different, belonging to Africa and the East. If Mycenae resembled the monuments of County Meath, the ancient civilization which had built it needed to be seen in terms of its 'familiarity with Egypt', a relationship demonstrated in such details as the lotus pattern discovered on a blade at Mycenae, or a palm motif which indicated 'that the general range of the civilisation was that of Africa'. Having asserted such differences between Greece and Ireland, Mahaffy qualified his remarks by claiming that Mycenaean culture also 'reached out to the north of Europe' and 'the silver-headed elk and reindeer or elk, found in grave IV., can only be the result of northern intercourse'. Reciprocally, Mahaffy added, citing evidence from recent excavations, 'we see in Celtic ornament the obvious reproduction of the decorations of Mycenae' (Mahaffy 1892, 426–8). Greece, no less than Ireland, was part of Europe, even while 'psychologically and pragmatically [it] partook of attitudes best called colonial' (Foster 1989, 163; see also Blaut 1993).

CONCLUSION: GREEK CARICATURES

The present chapter has focused on European (and in particular, British) travellers' endeavours to map Greece in and around the end of the nineteenth century. It has sought to show how Greece was construed, on the one hand, as a familiar European nation, and how, on the other hand, it was envisaged as an exotic Oriental country. These two oppositional models merged in an idea of familiar exoticism epitomized by colonial Ireland. Finally, larger concerns about the definitions of the European and the Oriental, or Asiatic, provide an important background to travellers' preoccupations with the location of modern Greek culture in relation to the national cultures championed by nationalist ideologies. An examination of travelogues to Greece demonstrates the extent to which nineteenth-century commentators wrestled with the fiction of a colonial other, and sought to erect watertight divisions between a secure 'us' and an unstable 'them'.

This essay, however, has *not* sought to explore the reciprocal concerns of Greeks with the place of Greek culture. The danger attendant on such a one-sided inquiry is that it runs the risk of promoting an artificial opposition between travelling and sedentary cultures (Clifford 1997, 17–46). Anxiety about such a one-sided enquiry was articulated by the author of a study entitled *Turkey in Europe*, which appeared in 1900. While accepting the validity of the West/East opposition, the book nevertheless inquired into the East's perceptions of the West. 'There remains, however, one point which it is important to elucidate,' declared the anonymous author, 'namely, what is the opinion of the Turks about the Eastern Question? What does the East think of the West?' Acknowledging 'the determination of Europe to impose its civilisation on uncivilised and half civilised nations all over the world', the author remarked that 'scant attention is paid to the large class of people who object to western civilisation' (Odysseus 1900, 3).

Greeks of all people were prodigious travellers and the second half of the nineteenth century witnessed a heated debate in Greece about the place of Greek culture, which was summarized in the title of an article published in 1842, 'What is Greece? Orient or Occident?' (Renieris 1842). Moreover, it is important to stress the extent to which Greeks themselves were shaped by and reciprocally shaped travellers' attitudes to Greece. Travellers' accounts of visits to Greece were translated into Greek, reviewed and debated in the Greek press: 'Serving as a mirror, a report card and an advertisement, travel writings often helped calibrate progress and the distance still to be travelled before Greece could join the company of the "civilized nations of Europe"' (Bastea 1997, 48). Murray's *Handbook* implicitly acknowledged this reciprocal process when it noted that 'The Greek seems to a foreigner perpetually playing the part of his own caricature' (*Handbook* 1884, 61). Greeks responded to representations of themselves in the European Press, just as British travellers arrived in Greece with their own preconceptions fed by Greek fiction and propaganda (Herzfeld 1995; Varouxakis 1992; Skopetea 1992). Indeed, Greek fiction itself enjoyed considerable popularity abroad and the final decade of the nineteenth century witnessed a spate of popular translations from Greek into English. An extreme form of this outsider's view of Greece, produced explicitly by a Greek for a non-Greek audience, was the children's writer Julia Dragoumis' volume, *Tales of a Greek Island*, published in English in 1912, before it appeared in Greece.

It was precisely this exotic, touristic literature for foreign consumption that the author and journalist Dimitrios Chatzopoulos condemned in an article published in the short-lived periodical *Dionysos* at the turn of the century, where he drew attention to the ways in which Greek writers were influenced by stereotypical projections of Greek rural life then fashionable in Europe. Many of the short stories, he complained, read like translations from foreign languages (Chatzopoulos 1901, 82–3).

While Greeks responded to European preoccupations with the folkloric, these demands on the part of an English readership for exotic material were reciprocally fuelled by precisely such writing. Chatzopoulos related how an Englishwoman arrived in central Athens and, to the consternation of a local inhabitant, expressed frustration at being unable to find a single brigand (Ibid., 83). As Chatzopoulos conceded, travel and translation were inseparable processes.

NOTES

1 See also the notice in *The Times*, 28 January 1880, p. 3, col. 5 and p. 4, cols. 3–5.
2 The publication of James Rennell Rodd's folkloric material assembled during his travels through Greece between 1888 and 1891 is another example of the folkloric interest of travellers (Rodd 1892).
3 For one of the first accounts of train travel in Greece, however, see Armstrong (1893).
4 In this article Clifford defends Townsend and refutes Bevan on the grounds that 'Fact, rigid and unbending – the everyday fact of Asia seen not through the smoke of the student's lamp, but through its native sun-glare – rises up to shatter the whole fabric of theory' (Clifford 1903, 129).
5 Michael Herzfeld has used Greek ethnography as a mirror for anthropology as a practice (1987).
6 As Larry Wolff (1994) has shown, this division of Europe along an East/West, as opposed to a South/North, axis was a late eighteenth-century development, influenced by the philosophers.
7 The French were the first to establish an archaeological school (1846), followed later in the century by the Germans (1868), the Americans (1882), the British (1885) and the Italians (1909).
8 On Lord Bute, a passionate philhellene, see Macrides (1992).
9 *The Times*, 24 October 1884, p. 8, col. 2.
10 Ramsay also took an interest in contemporary political developments in the area. See, for example, his article 'What to do in the East', *The Contemporary Review* LXXII (August 1897), pp. 234–41, and his account of the Young Turk revolution of 1908, *The Revolution in Constantinople and Turkey: A Diary* (London, 1909).
11 For a brilliant critique of Said, to which I am much indebted, see Ahmad (1992, 157–219).

12 As Elleke Boehmer notes, post-coloniality can be defined as 'that condition in which colonial peoples seek to take their place, forcibly or otherwise, as historical subjects' (Boehmer 1995, 3).

13 One recent study of the Greek Enlightenment, which draws upon the insights of Homi Bhabha and Partha Chatterjee, has sought to make comparisons between the experiences of India and Greece (Gourgouris 1996). Gourgouris's characterization of Greece as 'the paradigmatic colonialist condition' and Greece's conflation with India ('these two stories have a common history'), even with the proviso 'in decidedly different ways' (6–7), however, points to both the limitations and the usefulness of Greece's inclusion as a case study in ongoing discussions of the colonial and post-colonial. To speak of a 'paradigmatic condition', on the one hand, is to dismiss difference and historical specificity in favour of a single post-colonial narrative that essentializes both – as if Greece needed to become India or Egypt in order to be deemed worthy of critical interest. On the other hand, Greece offers an important case study precisely because the Greek experience undermines the singularity of the 'common story' of colonialism and its resistance as transhistorical and global experiences.

14 This was inextricably bound up with environmental issues about place and culture, as well as the authenticity of translations (Webb, 1982; Spencer 1954).

15 D'Estournelles was the author of *La Vie de Province en Grèce* (Paris: Hachette, 1878).

16 See also the article in the *Blackwood's Magazine* 108 (1870), pp. 240–5.

17 A. Proust spent the winter of 1857 in Athens.

18 In fact, as John Pemble notes, only in Greece and the Holy Land was travel largely reliant on horse power, whereas Sicily, Egypt and Algeria were all connected with main railway lines (1987, 29).

19 See the articles and correspondence in *The Times* in 1905: 10 January, p. 5, col. 4; 11 February, p. 5, col. 2; 25 February, p. 4, col. 4. Greece functions as an important topos in Woolf's late novel *Jacob's Room* (1922).

20 *The Times*, 19 January 1885, p. 6, col. 1.

21 *The Times*, 20 January 1885, p. 5, col. 6.

22 Jenkyns notes that 'it was a commonplace to say that the British Empire was really two empires, consisting on the one hand of the settlement colonies, in which the population was largely British in origin, on the other of subject peoples governed more or less despotically' (1980, 333).

23 Indeed, Greek antiquity was scanned as a prototype of the folklore of Ireland (Dorson 1966).

REFERENCES

Ahmad, A. (1992) *In Theory: Classes, nations, literatures*, London: Verso.

Anderson, B. (1983 [1991]) *Imagined Communities: Reflections on the origins and spread of nationalism*, London: Verso.

Armstrong, I. (1893) *Two Roving Englishwomen in Greece*, London: Sampson and Low.

Augustinos, O. (1994) *French Odysseys: Greece in French travel literature from the Renaissance to the Romantic era*, Baltimore: Johns Hopkins University Press.

Baedeker, K. (1889) *Greece. Handbook for travellers*, trans. 2nd German edition, Leipzig: Baedeker.

Bastea, E. (1997) 'Nineteenth-century travellers in Greek lands: politics, prejudice and poetry in Arcadia', *Dialogos* 4, pp. 47–69.

Belle, H. (1881) *Trois Années en Grèce*, Paris: Hachette.

Bent, J.T. (1886) 'A christening on Karpathos', *Macmillan's Magazine* 54, pp. 199–205.

—— (1891) 'Modern life and thought amongst the Greeks', in *National Life and Thought of the Various Nations throughout the World*, London: T. Fisher Unwin.

Bevan, E.R. (1903) 'Asia and Europe', *Monthly Review* 10 (29), February, pp. 62–80.

Blaut, J.M. (1993) *The Colonizer's View of the World: Geographical diffusion and Eurocentric history*, New York: Guilford.

Boehmer, E. (1995) *Colonial and Postcolonial Literature*, Oxford: Oxford University Press.

Boon, J.A. (1982) *Other Tribes, Other Scribes: Symbolic anthropology in the comparative study of culture*, Cambridge: Cambridge University Press.

Butor, M. (1974) 'Travel and writing', trans. J. Powers and K. Lisker, *Mosaic* 8, 1–16.

Buzard, J. (1993) *The Beaten Track: European tourism, literature, and the ways to 'culture' 1800–1918*, Oxford: The Clarendon Press.

Cardinal, R. (1997) 'Romantic travel', in R. Porter (ed.), *Rewriting the Self: Histories from the Renaissance to the present*, London: Routledge, pp. 135–55.

de Certeau, M. (1986) *Heterologies: Discourse on the Other*, trans. Brian Massumi, foreword by Wlad Godzich, Manchester: Manchester University Press.

Chatzopoulos, D. (Boem) (1901) 'Emis ke Meriki Xeni', *Dionysos* 1, pp. 82–9.

Clifford, H. (1903) 'The East and the West', *Monthly Review* 11 (31), April, pp. 128–42.

Clifford, J. (1988) *The Predicament of Culture: Twentieth-century ethnography, literature and art*, Cambridge, MA: Harvard University Press.

—— (1997) *Routes: Travel and translation in the late twentieth century*, Cambridge, MA: Harvard University Press.

Clogg, R. (1986) *A Short History of Modern Greece*, Cambridge: Cambridge University Press.

Curzon, G. (1907) *Frontiers*, Oxford: The Clarendon Press.

Deschamps, G. (1892) *La Grèce d'Aujourd'hui*, Paris: Armand Colin.

D'Estournelles, Baron P. (1889) 'The superstitions of modern Greece', *The Contemporary Review* 62, April, pp. 586–605.

Dorson, R.M. (1966) 'The question of folklore in a new nation', *Journal of the Folklore Institute* 3, pp. 277–98.

Ellman, R. (1987) *Oscar Wilde*, London: Hamish Hamilton.

Farrer, R.R. (1882) *A Tour in Greece (1880), with twenty-seven illustrations by Lord Windsor*, London and Edinburgh: William Blackwood.

Forster, E.M. (1908 [1986]) *A Room with a View*, Harmondsworth: Penguin.

Foster, R.F. (1989) *Modern Ireland 1600–1972*, Harmondsworth: Penguin.

Frazer, J.G. (1898) *Pausanias's Description of Greece*, London: Macmillan.

Gennadius, J. (1892) 'Preface' to *Kolokotrones. The Klepht and the warrior. Sixty years of peril and daring. An autobiography*, trans. Mrs E.M. Edmonds, London: T. Fisher Unwin, pp. v–xxv.

Gourgouris, S. (1996) *Dream Nation: Enlightenment, colonization and the institution of modern Greece*, Stanford: Stanford University Press.

A Hand-book for Travellers to the Ionian Islands, Greece, Turkey, Asia Minor, and Constantinople (1840) London: John Murray.

A Handbook for Travellers in Greece (1854) 2nd edition, London: John Murray.

A Handbook for Travellers in Greece (1872) 4th edition, London: John Murray.

A Handbook for Travellers in Greece (1884) 5th edition, London: John Murray.

A Handbook for Travellers in Greece (1896) 6th edition, London: John Murray.

Herzfeld, M. (1987) *Anthropology through the Looking-Glass: Critical ethnography in the margins of Europe*, Cambridge: Cambridge University Press.

—— (1995) 'Hellenism and Occidentalism: the permutations of performance in Greek bourgeois identity', in J.G. Carrier (ed.), *Occidentalism: Images of the West*, Oxford: The Clarendon Press, pp. 218–33.

Hillis Miller, J. (1995) 'Border crossings, translating theory: Ruth', in *Topographies*, Stanford: Stanford University Press.

Jebb, R.C. (1880 [1901]) *Modern Greece. Two lectures delivered before the Philosophical Institution of Edinburgh, with pages on 'The Progress of Greece' and 'Byron and Greece'*, London: Macmillan.

Jenkins, R. (1961) *The Dilessi Murders*, London: Longmans.

Jenkyns, R. (1980) *The Victorians and Ancient Greece*, Oxford: Blackwell.

Kiberd, D. (1996) *Inventing Ireland: The literature of the modern nation*, London: Vintage.

—— (1997) 'Modern Ireland: postcolonial or European', in S. Murray (ed.), *Not on Any Map: Essays on postcoloniality and cultural nationalism*, Exeter: University of Exeter Press, pp. 81–100.

de Lamartine, A. (1850) *Travels in the East, Including a Journey in the Holy Land*, vol. 1, Edinburgh: William and Robert Chambers.

Lee, H. (1996) *Virginia Woolf*, London: Chatto and Windus.

Lynch, E.M. (1900) 'Greece and Ireland', *The Gentleman's Magazine* 289, July, pp. 43–56.

Macrides, R.J. (1992) *The Scottish Connection in Byzantine and Modern Greek Studies*, St Andrews: Centre for Advanced Historical Studies.

Mahaffy, J.P. (1892) *Rambles and Studies in Greece*, revised 4th edition, London: Macmillan.

Nordau, M. (1895) *Degeneration*, London: William Heinemann.

Odysseus (1900) *Turkey in Europe*, London: Edward Arnold.

Pemble, J. (1987) *The Mediterranean Passion: Victorians and Edwardians in the South*, Oxford: The Clarendon Press.

Pick, D. (1996) *Faces of Degeneration: A European disorder, c. 1848–1918*, Cambridge: Cambridge University Press.

Playfair, R. L. (1881) *Handbook to the Mediterranean, its cities, coasts and islands for the use of general travellers and yachtsmen*, London: John Murray.

Pratt, L. (1992) *Imperial Eyes: Travel writing and transculturation*, London: Routledge.

Ramsay, W.M. (1890) *The Historical Geography of Asia Minor*, London: John Murray.

Renieris, M. (1842) 'Ti Ine i Ellas', *Eranistis* 2 (1), pp. 187–213.

Rodd, J.R. (1892) *The Customs and Lore of Modern Greece*, London: Stott.

Said, E. (1979) *Orientalism*, New York: Vintage.

Sergeant, L. (1880) *Greece*, London: Sampson Low, Marston, Searle and Rivington.

—— (1897) *Greece in the Nineteenth Century: A record of Hellenic emancipation, 1821–1897*, London: T. Fisher Unwin.

Skopetea, E. (1992) *I Dusi tis Anatolis: Ikones apo to telos tis Othomanikis Aftokratorias*, Athens: Gnosis.

Smith, A. (1884) *Glimpses of Greek Life and Scenery*, London: Hurst and Blackett.

Spencer, T. (1954) *Fair Greece, Sad Relic*, London: Weidenfeld and Nicolson.

Steevens, G.W. (1897) *With the Conquering Turk: Confessions of a Bashi-Bazouk*, London and Edinburgh: William Blackwood.

Super, R.H. (ed.) (1960–77) *The Complete Prose of Matthew Arnold*, vol. 1, Ann Arbor: University of Michigan Press.

Thackeray, W.P. (1844 [1991]) *Notes on a Journey from Cornhill to Grand Cairo*, Heathfield: Cockbird Press.

Todorova, M. (1997) *Imagining the Balkans*, New York: Oxford University Press.

Townsend, M.W. (1901) *Asia and Europe: Studies representing the conclusions formed by the author in a long life devoted to the subject of the relations between Asia and Europe*, Westminster: Archibald Constable.

Tozer, H.F. (1873) *Lectures on the Geography of Greece*, London: John Murray.

Turner, F.M. (1981) *The Greek Heritage in Victorian Britain*, New Haven: Yale University Press.

Upward, A. (1908) *The East End of Europe: The report of an unofficial mission to the European provinces of Turkey on the eve of the revolution*, London: John Murray.

Urquhart, D. (1838) *The Spirit of the East, illustrated in a journal of travels through Roumeli during an eventful period*, 2 vols, London: Henry Colburn.

Varouxakis, G. (1995) 'The idea of "Europe" in nineteenth-century Greek political thought', in P. Carabot (ed.), *Greece and Europe in the Modern Period: Aspects of a troubled relationship*, London: King's College London, pp. 16–37.

Webb, T. (1982) *English Romantic Hellenism, 1700–1824*, Manchester: Manchester University Press.

Weber, E. (1979) *From Peasants into Frenchmen: The modernization of rural France 1870–1914*, London: Chatto and Windus.

Wolff, L. (1994) *Inventing Eastern Europe: The map of civilization on the mind of the Enlightenment*, Stanford: Stanford University Press.

Woolf, V. (1906 [1987]) 'A dialogue upon Mount Pentelicus', in *The Times Literary Supplement*, ed. S.P. Rosenbaum, 11–17 September, p. 979.

Wyse, Sir T. (1871) *Impressions of Greece with an introduction by Miss Wyse, and letters from Greece to friends at home by Arthur Penrhyn Stanley*, London: Hurst and Blackett.

9 Travelling through the Closet

Michael Brown

INTRODUCTION

'[T]heories travel,' Edward Said (1983, 226) tells us. But to what end? He suggests their inevitable mobility (across disciplines, between material locations and amid different theorists) carries us to the frontiers and boundaries of theory's ability to explain. It is not surprising then that travel writing – or more accurately the geographies it reveals – might be used to confront the utilities and the limits of a particularly influential body of theory. I would like to goad such a travelling theory in this chapter, namely psychoanalytic accounts of desire, and more specifically their own travels into literary theory. I will argue that these accounts can be brought into conversation with written geographies to produce critical insights about the mechanics of desire, especially when it takes such a spatial textualization: *the closet*. This long-standing spatial metaphor signifies what Eve Sedgwick (1990) has deemed the fundamental architecture of gay oppression this century. The closet evokes a sense of concealment and erasure typical of lesbian and gay desire. So much so, in fact, that it has itself travelled to signal any denial or ignorance of one's identity. For instance, we now talk about people who are 'in the closet' about their HIV status when they deny their own seropositivity or refuse to tell others about it.

If the closet represents the place where gay and lesbian desire remains hidden, what sort of space is it? In this chapter, I am specifically interested in how the closet spatializes sexual desire for lesbians and gay men. The sign 'closet', I want to argue, is precisely such an articulation between sexuality, space and desire. Travel writing about the closet, then, provides a particularly appropriate venue for understanding their relations textually. I want to think through some of these relations by considering the travel writing of American gay author, Neil Miller. In his two books, *In Search of Gay America* (1989) and *Out in the World* (1992), Miller explores (quite literally) the geographies of the closet at national and global scales. More specifically, I will focus on his travels into two of the most closeted places on his tours: Selma, Alabama and Hong Kong. Though these are clearly very different spatializations of the closet, their differences highlight the range of ways the closet can work on desire. I will draw on two often competing foci on desire in psychoanalytic/literary theory – Lacan's and the schizoanalysis of

Deleuze and Guattari – to analyse the workings of the metaphor. Through Miller's travel writing I will show how psychoanalytic work on desire, when brought into contact with gay writing about space and desire, may launch us off into different (some might say opposing) directions *vis-à-vis* what is important about desire. That dissonance, however, worries me because it obscures the geographical point Miller makes so cogently: that where we desire enables and constrains how we desire.

PSYCHOANALYSIS AND LITERARY THEORY

The links between psychoanalytic theory and literary criticism are tangled and not always clear. And at first glance, the two intellectual pursuits seem interested in entirely different fields: the psychoanalyst is interested in the human unconscious; the literary critic is interested in the written text. There has been a substantial exchange of ideas between the two camps, however, though as Brooks (1987) points out, it has been largely one sided, with literary theory being largely the recipient of insights from psychoanalysis. The rationale for theoretical travel between these two domains can be understood through three closely related claims. First, scholars claim that the two inquiries have similar subject matter. Here we see Lacan's (1977, 147) famous claim that 'the unconscious is structured like a language'. Gallop (1985), for instance, argues the association is based upon an analogy: the literary critic is like the psychoanalyst. Likewise Wright (1984) advises critics to work on a text like a psychoanalyst would work on a psyche. Thus the parallels in the production of signs' meanings through metaphor and metonym can be drawn between a person's dreams and a piece of literature.

It follows, then, that if the subject matter is similar, so too are the tactics of investigation, and this is a second connection. Both literary criticism and psychoanalysis rely on the premise that there is a hidden level of meaning below the surface. What either a text or a human behaviour seems to mean at first glance is not necessarily what it does mean. In psychoanalysis, Freud's 'discovery' of the unconscious established this occluded world of signification. In literary criticism, Eagleton (1983) argues, the (often implicit) notion of a 'subtext' presumes exactly the same structure of meaning. Both pursuits therefore devise strategies of deepening or expanding their reading of things. Both search for contradictory or ironic truths that stand in tension to the more 'conscious' or 'literal' meanings at hand.

Finally, both their subject matters and their techniques illustrate what Selden and Widdowson (1993, 136) call 'the articulation of sexuality in language'. Questions about our relationship with desire, and the signification systems that we use to express or suppress it, are clearly bound up in both pursuits. So for instance, Eagleton (1983) suggests that there is a confluence between the two intellectual pursuits because both are interested in the same thing: how human beings pursue pleasure. Similarly, Brooks (1987, 4) claims their paths cross 'where literature and life converge'.

Psychoanalytic literary criticism, then, seems to be an appropriate starting point in considering the relations between desire, text and space in travel writing.[1] It enables an understanding of the way the closet metaphor works textually in travel writing. Conversely travel writing about the closet, like Miller's, would attest to the utility and insightfulness of this mode of criticism. Before turning to that encounter, however, it is necessary to consider how the spatial metaphor of the closet actually works.

THE CLOSET

When we use the metaphor of the closet we are *geographically* portraying the strong textual link that is made between identity and desire. In other words, who we are has a great deal to do with whom or what we want subconsciously. And I am intrigued by the closet metaphor because the link it makes is so decidedly spatial. Metaphors work through the substitution of signs based on similarity, literary theory tells us (Jakobson 1990; Eagleton 1983). This capacity for substitution, however, does not mean there is a precise equivalence between the two signs. What allows a metaphor to work, then, also enables it to fail: the fact that the exchange of signs is based on recognizable similarities as well as differences (Swanson 1978). So as a geographer I am interested in what alleged qualities of space enable an exchange through the trope of the closet.

While the closet is such a commonplace metaphor, its etymological origins are unknown. To date there has been no genealogy or social history of 'the closet' as a term of desire, yet lesbian and gay studies have been arduously exploring the historical geographies it signifies (e.g. Lapovsky-Kennedy and Davis 1993; Chauncey 1994). The goal for most of these social histories has been to represent the social worlds of gay people before Stonewall. They present a record and give a sense of what specific closets were like in history. Thus their goal is to render the closet rightfully visible. By establishing such a presence, however, these studies strengthen the taken-for-granted conceptual links that are made between sexuality, desire and space in the metaphor.

Why has it become so common, so vernacular, to represent marginal sexuality spatially? At first glance the answer seems obvious. A closet is a bounded physical space and the things inside it are not immediately visible. It denotes a place where you hide things (or yourself) (e.g. Jay and Young 1972; Gross 1993; Signorile 1993). You 'put things away' there. Placed in a closet, items are not immediately relevant or useful. They are often forgotten there – 'Out of sight, out of mind.' At the very least, things inside the closet lack an immediate presence. The closet signifies a confining, fixed space, one that decidedly obscures its contents. The spatial metaphor of the closet is an appropriate venue to explore the geography of same-sex desire for two reasons. First, it highlights the fact that desire has a materiality, a 'whereness' that is often elided when literary theory reads psychoanalytic theory. Thus the closet demarcates or maps desires and their flows between subjects. Secondly, the closet has come to constitute a particular form of desire: same-sex sexual desire in the twentieth century. It suggests that the form desire takes can, as it were, act back on that desire itself. In other words the closet has shaped the ways gays and lesbians desire. What happens to our thinking about desire, then, when we consider the material spaces where gays and lesbians actually live, where they actually do their desiring? To phrase it more academically, what happens when we spatialize 'the closet'?

Miller's geography of the closet

Neil Miller is not a scholar *per se*; he is a journalist by trade. He was editor of Boston's *Gay Community News* in the mid-1970s and more recently has written for *The Boston Phoenix*, an entertainment and arts weekly. Consequently, his self-professed aim with these travelogues was quite descriptive. As he put it, 'I didn't start out with any particular preconceptions, any grand theory I wanted to prove. My intention was to paint up-close portraits of people and communities, letting my subjects tell their own stories' (Miller 1989, xv). Still, these books do have theories travelling through them: theories about space and desire. The impetus for the books came in the mid-1980s, when he noticed the extent and speed with which AIDS had forced people out of the closet, a process Dennis Altman (1988) has referred to poignantly as 'legitimation through disaster'. The AIDS crisis propelled people out of the closet on a number of scales. It forced many people living with AIDS to disclose their sexuality publicly. As it hit gay men through the 1980s, their communities became much more visible as they fashioned responses to the epidemic and demanded that others take notice of the holocaust in their midst (Brown 1994). Miller watched these trends avidly. He witnessed a 'moving in from the margins' unfold in his native Boston, as well as in other urban gay centres like New York, London and San Francisco (Miller 1989, 9). The closet metaphor seemed to be failing him in these urban texts. Gays and lesbians were becoming more visible; they were gaining an explicit, widely recognized presence in places. Yet simultaneously, he admitted, it was as apt as ever. 'Some things just didn't change.' A rise in anti-gay violence could be documented by the mid-1980s. It was still considered acceptable to be anti-gay in public discourse. And even 'liberal' Massachusetts explicitly barred lesbians and gays from being foster-parents. The paradox was enough to send Miller packing. He decided to investigate this duality and document it through travel. He toured small towns across the United States, places like Selma, Alabama, Morgantown, West Virginia, and Fargo, North Dakota. In these locales he sought out lesbians and gay men to talk about their lives, their positions in and out of the closet.

> I wanted to know, 20 years after the Stonewall Riots, if gay pride and progress had finally begun to trickle down to the grassroots; if the options for living, working, and being belligerently ourselves that we had won in cities like San Francisco, New York, and Boston extended to the rest of the country, to the towns where so many of us had grown up, to minority communities, to mid-size cities. I wondered too, as a gay man living in a comfortable urban

environment, if I might have something to learn – about survival, about community and finding one's own place in society – from people in less congenial places.

(Miller 1989, 11)

A few years later he undertook a more global expedition of lesbian and gay life, again to chronicle its existence inside and out of the closet. The places he visited included South Africa, Egypt, Hong Kong, Thailand, Japan, Argentina, Uruguay, Australia and New Zealand. While his books are by no means comprehensive or even entirely representative of gay and lesbian life, they nonetheless provide a certain in-depth-yet-extensive cartography of gay and lesbian identities in places. They show how sexual identity and desire are geographically mediated.

Miller's conclusions at the end of his travels perhaps come as no surprise to any critical human geographer. He finds no common theme, no patterned geography of desire, no model architecture of the closet. In his own words, 'I found no single vision of gay future. Instead I saw various gay and lesbian communities taking different paths, adopting their own strategies as they attempted to create an atmosphere of openness and a sense of security' (Miller 1989, 306). Substantively interesting as his travel writing is, I think the plurality it represents has important theoretical implications for psychoanalytic theories of desire in literary work, and the metaphor of the closet in particular. Through his work, I find myself tracking theories' (in this instance, of desire) travels between the material closets inhabited by desiring lesbians and gays and the metaphorical closets of desire built through psychoanalysis. Miller's travel writing is nothing less than a geography of the closet. It charts where and how gay and lesbian desire takes place. In order to trace this route, it is necessary to sketch out the ways that psychoanalysis has theorized desire itself.

GEOGRAPHIES OF DESIRE

There are many geographies to take from Miller's work. For instance, his travels demonstrate what Luke (1994) has termed 'global' politics, as struggles to create livable local gay space can have effects at a variety of other spatial scales. It also strikes me that Miller's travel writing has much to inspire the recent spate of comparative work in urban politics, and attempts to sensitize them to scale issues. Another important geography Miller self-consciously explores is the negotiated one between lesbians and gay men. He is convinced that a viable politics of sexuality must include both gays and lesbians. Thus this travel writing pays close attention to the coalition-building work done to overcome the so-called 'limits' of identity politics that have often split sexual politics along gender lines. But given my interest in the spatiality of the closet, I am most struck with the way Miller's work mediates psychoanalytic theories of desire spatially. I will specifically focus on the debate between Jacques Lacan's (1973; 1977) Oedipally centred theory of desire and Deleuze and Guattari's (1972) 'schizophrenic' account. Both of these long-standing perspectives continue to inform psychoanalysis's take on how desire works, and their insights on that process have also 'travelled' towards other disciplines, such as literary theory (Eagleton 1983; Wright 1984; Selden and Widdowson 1993).[2]

Before I sketch out the terrain of debate between Lacan and Deleuze, it helps to understand their polarity by identifying the insights and premises they share. As poststructuralists, both reject a unified theory of the subject.[3] While each accepts a necessarily decentred subjectivity, they nonetheless also agree that subjectivity and desire are at the very least completely bound up with one another. Both are working against Freud's more biological explanations for human desire that pervade psychoanalysis, especially those that source desire in some innate instinct. Finally, both aspire to portray desire much more diffusely than simply placing it all upon a specific object of desire.

Despite these shared beginnings, however, the two theorists develop ideas about desire that are quite at odds with each other. Typically, they are portrayed as poles in a debate about how desire works (e.g. Hocquenghem 1972; Bogue 1989; Selden and Widdowson 1993; Grosz 1990; Fuery 1995). Hocquenghem (1972) in particular uses Deleuze's work specifically to criticize Lacan, whom he positions as largely a Freudian progeny. Bogue (1989, 3) argues more specifically that Deleuze had long been hostile to the implicit Hegelianism in Lacan's theory of desire.[4] He first critiqued it in *Nietzsche and Philosophy* (1962). *Anti-Oedipus* is a widening of that rift with Lacan, albeit an explicitly politicized one given Guattari's activism. This seminal book is a trenchant critique of Freudian insistence on the Oedipal origins and mouldings of human desire. While Lacan, too, draws away from Freud in important respects (see Bowie 1979; Grosz 1990), his work clearly extends and reinforces an Oedipally centred human desire.

Lacan and desire

Let me now sketch this debate's presence in Miller's geography of the closet. For Lacan, when we desire we are trying to assert a sense of being. It is a way to mark our existence. But because of the Oedipus conflict, and our attempts to resolve it through the mirror phase, desire can never actually be satisfied. It can never actually be vanquished because of its otherness for (and through) social subjects. Desire is other according to Lacan because it shows subjects that their needs cannot be self satisfied; they must look elsewhere, outside the conscious self. It is perhaps less important that we wade through Lacan's vexing rhetoric and logic than focus on his portrayal of desire itself. In the Lacanian view, desire is always a *lack* for the subject. It is never satisfied; it never can be, but is constantly deferred. It always exceeds discrete desired objects themselves. Thus desire is a metonym, a sort of horizontal substitution of signifiers (Lacan 1977, 167). This is why he makes the well-known claim that the unconscious is structured like language. Given that subjectivity and desire are so closely bound up with each other for Lacan, we are said to be defined by our desire. In other words subjects are constituted by what they lack – not through any whole or unified subjectivity (which is the conventional model of humanity in Western philosophy). Desiring subjects are therefore marked by uncertainty, fragmentation and alienation (Grosz 1990; Selden and Widdowson 1993).

What does all this have to do with travel writing? Recall that I am interested in the allegedly spatial qualities that allow the closet to work as a metaphor. Lacan is certain that subjectivity is defined by a lack, by uncertainty, fragmentation and alienation. These features also highlight certain material aspects of the closet in Miller's geography. They can be traced through Miller's description of gay and lesbian life in Selma, Alabama.

While Selma is a medium-sized American city, its location in the deep South makes it a culturally conservative and generally homophobic place. Gay and lesbian desire there is clearly structured by the closet. Listen, for instance, to how Miller describes what the closet is like for 'Jill', a forty-year-old white lesbian he met in Selma, Alabama:

> Like other gay people she knew in Selma, Jill was in the closet. None of her neighbors were aware that she was a lesbian, except for one woman who was her best friend. She hadn't told anyone at her job, and she never set foot in the gay bar in Montgomery, out of fear of exposure.
>
> There were no gay bars in Selma, no gay, lesbian, or feminist groups, and no gathering place except Skeeters, the bar in Montgomery 58 miles away. 'I'm not sure there could ever be a place here in Selma,' said Jill. 'Maybe it's paranoia. But I think it would not be tolerated.'
>
> (Miller 1989, 18–19)

She went on to say: 'It's not even like the gay grapevine here because people are so far back in the closet, they are back behind several rows of clothes. Maybe underneath the shoes.'

From these passages we can heuristically link up the physical space of the closet with Lacan's account of desire. A closet is typically separate from the room it adjoins. It is at the margins of immediate space. In Selma desire is surely fragmented. It is kept separate from the immediate lifeworld. It is kept at bay not only from the public sphere, but the private sphere as well. To be in a closet physically is to stand apart from, but still inside, the room where the closet is located. It is to be alone. In Selma the alienation gays and lesbians feel is intense, even where gays know one another. Jill says they are 'back behind the shoes'. For Lacan, alienation is also a symptom of desire, which can never be sated. Standing apart from others in an actual closet, one cannot always be sure what goes on in the room itself. It is hard to see the centre from the margins. In Selma, Jill reflects that she might be paranoid, fearing how straights might react to same-sex desire there. But she can never really be sure. For Lacan, uncertainty always surrounds the desiring subject. We can read uncertainty, fragmentation and alienation in the imagery of the actual closet, just as we can perceive them in the intersection of subjectivity and desire of gays and lesbians in Selma, Alabama.

The themes of uncertainty, fragmentation and alienation are also manifest in Miller's discussion of Hong Kong in the early 1990s. Officially speaking, homosexuality does not exist in mainland communist China. Same-sex desire is depicted as a Western corruption. Yet as a British colony, in the early 1990s Hong Kong still had nineteenth-century anti-sodomy laws on the books. Cultural and legal structures have combined to construct a closet that strives to discipline, deny and erase same-sex desire. Thus in the colony, the closet works in several ways to promote the discourse of lack around desire. Most obviously, Miller describes how it was impossible for him to meet any lesbians in Hong Kong. They were too closeted for him to reach. Also, he goes on at length about how the closet in Hong Kong perpetuates the absence of a strong gay and lesbian community. He describes how gays and lesbians are starved of any sort of gay culture because of the ubiquity of the closet, and suggests this lack leads to a profound sense of uncertainty about who they are. Perhaps rather ironically, he discusses how gay and lesbian desire within the closet leads to a dangerous lack of anonymity. The need to preserve one's anonymity in Hong Kong, Miller shows, leads to a profoundly fragmented and alienated gay or lesbian subject. As one of his informants confided:

> Your friends are all around you. Your colleagues are very close to you. If something is known, the news passes quickly. Mostly, people are afraid of their sexual orientation being known, so they try not to go to the disco or the bar. These places are known to outsiders. So you meet others at the public toilet, instead. A quickie. But what about mental need? The spiritual need? You just keep it inside your heart. Closet! That's the word.
>
> (Miller 1989, 100)

Later, another informant, a writer who was daringly open about his sexuality, went on to describe just how powerful an effect the closet had on the relationships between people's identity and their desires:

> [H]e viewed contemporary Hong Kong as a big closet. 'We are geographically blocked,' he said. 'There is simply nowhere to be anonymous. You can't drive to Shanghai like you could in the 1930s.' Some of the symptoms of the closet were letters he had received after his books were published. One reader wrote . . . using his left hand, so his script couldn't be identified. Another letter was composed of Chinese characters cut out of newspaper, for the same reason.
>
> (Ibid., 104)

The metaphor of the closet has material referents that sustain the defining elements of Lacanian desire quite consistently. The closet perpetuates a lack around desire through tropes of alienation, uncertainty and fragmentation. We can imagine this lack as being confined in a closet, as never being able to stand in the actual room. Consequently, desire is always concealed: marginalized and peripheral.

Deleuze and Guattari and desire

Reading the closet this way seems to be at odds with a Deleuzian theory of desire. And yet such an alternative framework produces just as plausible and helpful a reading/rendering of the closet. Deleuze and Guattari have provided a strong critique of both Lacan and Freud. Their central argument against these early psychoanalysts hones in on the notion that desire is based in lack or need, which implies some sort of essential deficiency in human subjectivity. They decidedly reject the Oedipus complex as the pan-historical, ubiquitous law that defines all human desire in terms of a deficit. That representation of desire Deleuze and Guattari attribute as a function of capitalism. It is a tactic of interpellative, hegemonic control at the most micro scale. Writing amid the heady French (post)structuralism of the late 1960s, they insist that the mode of production 'deforms' the unconscious, a rather provocative extension of Freud's thinking about the social structuring of human consciousness. But parting with Freud, they argue that psychoanalysis cannot cure but only exacerbates things. The Oedipus complex, they insist, is a result of capitalism's general repression and channelling of a rather effusive human desire. Capitalist social relations, working through the institution of the modern family, repress desire.

Their alternative approach, labelled 'schizoanalysis', constructs an un-conscious in which desire is 'an untrammeled flow' (Selden and Widdowson 1993, 143). There is no single, correct form or direction desire will take. From this premise, then, there is no such thing as misplaced desire (which the Oedi-pal theory assumes). In turn, desire need not signal any necessary lack or deficit in the human subject. 'Desire isn't a lack,' they argue, 'it's just desire.' Rather than being contained by Oedipal anxiety, desire is conceptualized as an energy, a positive source for new beginnings, a 'voyage of discovery'. Rethinking desire in this way, Deleuze and Guattari call it productive, creative, generative and multiple (1972). Through schizoanalysis, they seek to liberate desire from the limits and stigma placed on it by capitalism and exacerbated by modern psy-choanalysis. They call this strategy 'deterritorialization', suggesting mobility instead of fixity. They want to liberate desire from oppressive institutions (or 'territories') of the family, church, nation, school and, of course, psychoanalysis itself.

Deleuze and Guattari's thinking has had some appeal in queer theory. Hocquenghem (1972), for instance, has drawn extensively on their work to expose the heterosexism that saturates both Freud's and Lacan's work on desire. He (like many queer thinkers and activists influenced by the sexual revolution) stresses the important *political* point that gays and lesbians should not necessarily import definitions of desire from heterosexuality, but work creatively to produce their own (see Calfia 1995). Hocquenghem shows that the stakes of a singular Lacanian reading of queer desire are high indeed!

Reading Miller's travels with this more generative perspective on desire provides a rather different – but no less accurate – geography of the closet. I want to be clear about what I am suggesting here: I am not using schizoanalysis to argue that there were smaller or larger closets across different places. Nor am I saying that there were more people in the closet in some of the places Miller visited than in others. It is simply that when I think about the geographies of the closet I find that a Lacanian reading of desire seems to deflect attention away from the strategies lesbians and gays have used to desire – in spite of, or in opposition to, the closet. This is a powerful motif in Miller's writing. So, for instance, in Hong Kong, despite the thorough ubiquity and intensity of the closet, Miller admits that the closet did, in fact, have a geography: the bar, the disco, the public lavatory. In other words, there were spaces productively carved out within the closet where desire could be pursued. More obliquely, in Selma, despite the fact that there was no discrete gay space in the city, people could still make contact with others who shared their same-sex desires.

> Even in a town like Selma gays have an uncanny ability to make contact. 'You know the expression, it takes one to know one?' asked Jill. 'You just *know*, and once other gay people realize you are gay, they will make comments and you get to talking and they tell you things.' But all this was terribly discreet.
>
> (Miller 1989, 18)

Jill herself had recently been romantically involved with a woman from Birmingham and while she would often commute there, the couple would often spend the weekend at Jill's house in Selma, with her neighbours oblivious to the fact.

Yet another strategy was to feign heterosexuality by being in a heterosexual relationship and/or getting married, and then having affairs on the side. Miller documents the pervasiveness of this strategy in both locations. For example, he describes a chance encounter with a young gay man in Hong Kong:

> As we started back down the hill, he began talking rapidly and excitedly. He had been living a double life, he said, trying to hide from his family, his friends, and, above all, from his girlfriend. At work, he made sure to avoid gay people, who might have seen him near the bar or the disco; he feared that talking to them might give him away. He couldn't introduce his gay friends to his girlfriend. 'What would I say? "This is my Canadian friend. This is my New Zealand friend." What would she think?'
>
> (Miller 1992, 106)

In other words, just because people were (desiring) in the closet did not mean that they did not resist and adapt to the situation. Desire produced strategies of resistance because of the closet.[5] A much more expansive map of desire folds in *this* picture of the closet too.

Finally, the very presence of his informants in these places suggests a certain degree of resistance emerging from the productive, generative aspects of desire. A consistent theme in Miller's work is how creative gay and lesbian people have become in pursuing their sexual desires in the context of a homophobic closet. This suggests a certain productive dimension to closeted desire that may ultimately act back upon the constraints it imposes.

Now clearly I have not discussed the obvious constraints that limit these people's ability to pursue their desire creatively. Nor have I discussed the ethics or dangers of their chosen strategies. But that is because my point is not to argue the merits of these tactics. It is instead to offer a different geographical reading of desire, one that is informed by Deleuze and Guattari's schizoanalysis. In a closeted place (or a place of closets?) like Selma or Hong Kong desire was being 'deterritorialized'. It was, as Miller discovered, transgressing the limits and fixity placed upon it by the family, the church and the community. Gays and lesbians were actively liberating their desire. Recall that schizoanalysis portrays desire as productive, creative, generative and multiple. Each of these qualities might be used to describe the interior of the closet in Miller's travel writing. People devised several interesting and successful ways to pursue desire. Miller himself seems to recognize the power of schizoanalysis when he reflects on his travels through the closeted small towns of America. He is awed by the variability, strength and successes of their people:

> What struck me once again was this movement towards a richer, fuller life. . . . You couldn't have a one-dimensional existence if you lived in this part of the country. . . . Gays in small towns and rural areas had traditionally had those community ties and involvement. Usually they had been closeted – often in the deepest closet of all, a heterosexual marriage. Now many were cautiously trying to combine an allegiance to their roots with some of the openness and community that gays had in big cities. And so many people I met in these smaller towns seemed to have gone through enormous changes within the last six or seven years. There was still a fear of exposure, still homophobia of course. AIDS, while seemingly remote, heightened prejudice and suspicion but also enhanced the possibility of forging links with other gay people.
>
> (Miller 1989, 105)

It seems to me that we risk missing an important political point if we only take a Lacanian reading of desire when we spatialize desire through 'the closet'.

CONCLUSION

The purpose of this paper has been to use travel writing to prod a travelling theory. I have used psychoanalytic theory's migration into literary criticism to unpack the spatial metaphor of the closet. It suggests that the closet is a spatial metaphor used to portray gay and lesbian desire. Different takes on how to theorize desire, however, have produced different insights into the workings of the closet in a particular opus of travel writing. Lacanian theory points us towards the function of concealment at work in the metaphor. Located within the closet, lesbian and gay desire can be difficult for anyone, gay or straight (even the subjects themselves!), to find. Alternately, schizoanalysis draws our attentions toward the productive capacities of desire precisely because of their confinement as deviant and out of place. The closet does not necessarily alienate people from desire, or necessarily prevent them from translating it into wants, demands or wishes. I would like to close this essay by drawing out two implications for psychoanalytic literary criticism when the closet is spatialized, as Miller's travel writing has done so effectively.

First, Miller's in-depth/expansive geographies of lesbian and gay life paint a thoroughly equivocal picture of the closet for psychoanalytic theories of desire. It can be coded through a Lacanian take on desire, where the closet is a space of lack. It is a place of constant deferment and never-sated desire. It is often a place where people are striving to make themselves more complete but will always fail. But it can also be coded as a tremendously creative and productive space of desire. A schizoanalytic reading of the closet is just as plausible. Closets in material space were bursting with ingenious and productive strategies of desire. I get a much stronger sense of the equivocation of desire in Miller's work than I do from psychoanalytic texts. So more specifically what I wish to argue is that for all its poststructural rejection of dualisms and categorical imperatives, maybe psychoanalysis has some important geography lessons to learn about bothness, duality and betweenness in placing desire and describing its workings. By approaching desire through the closets in Miller's travels, we can see how a positioned academic debate occludes the simultaneity opposing perspectives can share in place. Returning to Said's axiom that opened this paper, it is not so much that theories travel, but that theories are always travelling *and* staying fixed in place.

Secondly, I would close by reflecting on what we should make of this closet of desire. Criticizing it as too static (as a number of geographers have suggested for spatial metaphors more generally) risks downplaying the vital political point that fixity is precisely how the closet can conceal and trap gays and lesbians (Brown 1996). The static nature of the metaphor points up a significant oppression. But on the other hand, stressing the limitations the closet places on desire draws attention away from a just-as-important – if antithetical – political point: that lesbians and gays in the closet can and do desire successfully through a number of creative and generative strategies.

One might argue that, as a geographer, I have fundamentally misread psychoanalytic/literary theory. That may be true; but if I have, I think the point about travelling theory which opened this paper becomes all the more salient. It is no longer done just by closed-off academic communities (of say, literary critics, or psychoanalysists). Consequently, when such discourses adopt spatial metaphors, they must be willing to import geographic theories about the dynamics of space and place.

I think the most useful way to approach the closet, then, is as a 'translation term' between sexuality, space and desire. As Clifford (1992, 110) explains, this is 'a word of apparently general application used for comparison in a strategic way'. He hastens to add to this definition: 'all translations used in global comparisons . . . get us some distance *and* fall apart' (original emphasis). The closet gets us (and the theories through which we travel) some distance in understanding the confinement and erasure of lesbian and gay desire, but it collapses when we approach the creative and innovative productions of desire people have pursued in places/closets simultaneously.

Acknowledgements

Thanks to Neil Miller for our discussions and for our travels in search of the perfect coffee across Cambridge and Somerville. A previous draft of this paper was given at the 1996 meeting of the Institute of British Geographers, Glasgow. I would like to thank Alison Blunt, Steve Pile and David Conradson for comments on previous drafts. A faculty research grant from the University of Canterbury funded this research.

NOTES

1 The relationship between psychoanalysis and queer subjects, however, has been by no means unproblematic (e.g. Bayer 1981; Lewes 1988). One might therefore view this essay as an attempt by queer theory to speak back to psychoanalysis.
2 Psychoanalysis, of course, has also travelled into geography, and several geographers are interested in a return trip. For some recent examples see Pile (1996), Pile and Thrift (1995), Rose (1993), Bondi (1993).
3 On the debate about whether Lacan can be 'properly' called a poststructuralist, see Leader and Groves (1995, 78). By labelling him as such, I am signalling his acknowledgement of structures (the autonomy of the symbolic) being paralleled by his insistence on retaining a place for the subject.
4 Fuery (1995, 14–15) notes that the specific influence is in Lacan's use of the master–slave model from Hegel. See also Sarup (1989, 21–2).
5 This is perhaps Chauncey's (1994) point in arguing against the metaphor of the closet in favour of the metaphor of a 'gay world' to describe New York in the early twentieth century (see Brown 1996).

REFERENCES

Altman, D. (1988) 'Legitimation through disaster: AIDS and the gay movement', in E. Fee and D. Fox (eds), *AIDS: the Burdens of History*, Berkeley: University of California, pp. 301–15.
Bayer, R. (1981) *Homosexuality and American Psychiatry: The politics of diagnosis*, Princeton: Princeton University Press.
Bogue, R. (1989) *Deleuze and Guattari*, London: Routledge.

Bondi, L. (1993) 'Locating identity politics', in M. Keith and S. Pile (eds), *Place and the Politics of Identity*, London: Routledge, pp. 84–101.

Bowie, M. (1979) 'Jacques Lacan', in J. Sturrock (ed.), *Structuralism and Since*, Oxford: Oxford University Press, pp. 116–53.

Brooks, P. (1987) 'The idea of a psychoanalytic literary criticism', in S. Rimmon-Kenan (ed.), *Discourse in Psychoanalysis and Literature*, New York: Methuen, pp. 1–18.

Brown, M. (1994) 'The work of city politics: citizenship through employment in the local response to AIDS', *Environment and Planning A* 26, pp. 873–94.

—— (1996) 'Closet geography', *Society and Space* 14, pp. 762–70.

Calfia, P. (1995) *Public Sex: The culture of radical sex*, Pittsburgh: Cleis Press.

Chauncey, G. (1994) *Gay New York*, New York: Basic Books.

Clifford, J. (1992) 'Travelling cultures', in L. Grassberg, C. Nelson and P. Treichler (eds), *Cultural Studies*, London: Routledge, pp. 96–116.

Deleuze, E. (1962 [1983]) *Nietzsche and Philosophy*, trans. H. Tomlinson, Minneapolis: University of Minnesota Press.

Deleuze, G. and Guattari, F. (1972 [1983]) *Anti-Oedipus: Capitalism and schizophrenia*, trans. R. Hurley, M. Seem and H. Rane, Minneapolis: University of Minnesota Press.

Eagleton, T. (1983) *Literary Theory: An introduction*, Minneapolis: University of Minnesota Press.

Fuery, P. (1995) *Theories of Desire*, Melbourne: Melbourne University Press.

Gallop, J. (1985) *Reading Lacan*, Ithaca, NY: Cornell University Press.

Gross, L. (1993) *Contested Closets: The politics and ethics of outing*, Minneapolis: University of Minnesota Press.

Grosz, E. (1990) *Jacques Lacan: A feminist introduction*, London: Routledge.

Hocquenghem, G. (1972 [1993]) *Homosexual Desire*, trans. Daniella Dangoor, Durham, NC: Duke University Press.

Jakobson, R. (1990) 'Two aspects of language and two types of aphasic disturbances', in L.R. Waugh and M. Monville-Burston (eds), *On Language*, Cambridge, MA: Harvard University Press, pp. 115–33.

Jay, K. and Young, A. (1972) *Out of the Closets: Voices of gay liberation*, New York: Douglas.

Lacan, J. (1973 [1994]) *The Four Fundamental Concepts of Psychoanalysis*, trans. A. Sheridan, Harmondsworth: Penguin.

—— (1977) *Écrits: A Selection*, trans. A. Sheridan, New York: Norton.

Lapovsky-Kennedy, E. and Davis, M. (1993) *Boots of Leather, Slippers of Gold: The history of a lesbian community*, New York: Routledge.

Leader, D. and Groves, J. (1995) *Lacan for Beginners*, London: Icon.

Lewes, K. (1988) *The Psychoanalytic Theory of Male Homosexuality*, New York: Meridian.

Luke, T. (1994) 'Placing power/siting space: the politics of local and global in the New World Order', *Society and Space* 12, pp. 613–28.

Miller, N. (1989) *In Search of Gay America: Women and men in a time of change*, New York: Harper and Row.

—— (1992) *Out in the World: Gay and lesbian life from Buenos Aires to Bangkok*, New York: Vintage.

Pile, S. (1996) *The Body and the City*, London: Routledge.

Pile, S. and Thrift, N. (1995) *Mapping the Subject: Geographies of cultural transformation*, London: Routledge.

Rose, G. (1993) *Feminism and Geography: The limits of geographical knowledge*, Cambridge: Polity Press.

Said, E. (1983) *The Word, the Text, and the Critic*, Cambridge, MA: Harvard University Press.

Sarup, M. (1989) *An Introductory Guide to Post-Structuralism and Postmodernism*, Sydney: Harvester Wheatsheaf.

Sedgwick, E. (1990) *Epistemology of the Closet*, Berkeley: University of California Press.

Selden, R. and Widdowson, P. (1993) *A Reader's Guide to Contemporary Literary Theory*, 3rd edition, Lexington: University of Kentucky Press.

Signorile, M. (1993) *Queer in America: Sex, the media, and the closets of power*, New York: Random House.

Swanson, D. (1978) 'Toward a psychology of metaphor', in S. Sacks (ed.), *On Metaphor*, Chicago: University of Chicago Press, pp. 161–4.

Willbern, D. (1989) 'Reading after Freud', in G.D. Atkins and L. Morrow (eds), *Contemporary Literary Theory*, Amherst: University of Massachusetts Press, pp. 158–79.

Wright, E. (1984) *Psychoanalytic Criticism: Theory and practice*, London: Routledge.

10 Writing over the Map of Provence

The touristic therapy of *A Year in Provence*

Joanne P. Sharp

Recently in the West there has been much lamenting of the demise of cultural difference, a public dismay over the closure of the Age of Exploration and the initiation of an age of homogenization. Despite this fear, the phenomenon of travel writing – a literary form apparently dependent upon difference and therefore doomed by its disappearance – is as popular as ever. Indeed, there appears to be something of a 'boom' in travel writing, especially in contemporary English society. Rarely have there been so many television programmes dedicated to all aspects of travel and tourism, from the cheapest package tour to the most adventurous 'rough guides' and expensive tailor-made affairs. Print culture also exhibits the English love of travel, with glossy colour supplements in the Sunday papers offering images of exoticism and romance.

Of all recent popular travel accounts by English writers, perhaps Peter Mayle's depictions of his move to and socialization into Provençal culture have been the most commercially successful. Although his account is of a less modernized way of life than most of his English audience would have experienced, his is not a narration that relies upon a naïve vision of unchanging rural ways. There are images of change and modernity in his accounts but, whereas he acknowledges the potential danger of these external influences upon the way of life he so admires, unlike other travellers, particularly anthropologists, he does not see the loss of local difference as being in any way certain. For Mayle, Provence engages with modernity in its own way to produce idiosyncratic, rather than homogenizing, results. This allows Mayle to go so far as to highlight the advantages of tourism and change when managed well: he contrasts his appropriate stance towards Provence against caricatured figures of culturally insensitive mass tourists whose influence he does consider to pose a potential threat. His is thus a therapeutic discourse that eases the anxiety of people self-conscious of their impact as travellers, giving them an assurance that local character and distinctiveness can – and will – be maintained in the face of global flows of tourism.

However, this therapeutic discourse is couched within a 'paradox of tourism': although Mayle's account self-consciously presents an image of tourism that does not destroy the cultural integrity of Provence, his books' popularity have led some to fear that his work has begun to change the imaginary and physical landscape, as more people seek the place and the experience that he describes.

WRITING THE ENGLISH COUNTRY IDYLL IN RURAL FRANCE

Peter Mayle's narration of his move to rural France, *A Year in Provence* (1989), and its sequel, *Toujours Provence* (1991), are arguably the most popular and commercially successful English-language travel writing of recent years. By 1992 *A Year in Provence* had topped the British paperback best-seller list for sixty weeks, and had been a best-seller for two and a half years (Aldridge 1995). *Toujours Provence* was also highly commercially successful. Nearly half a million paperback copies of the two books were sold in Britain alone in 1992, and in total, around 4 million copies have been sold worldwide. The popularity of Mayle's work generated a mini 'Provence-industry' in Britain which involved the re-release of Lady Fortescue's 1935 *Perfume from Provence* in 1992, sold as 'the original *Year in Provence*'; a satire, *A Year near Proxima Centauri*, which parodied Mayle's apparent obsession with consumption by setting the story on a planet where everything, including its residents, was potentially edible; a BBC TV adaptation starring popular English actors John Thaw and Lindsay Duncan; a semi-autobiographical 'novel of Provence', *Hotel Pastis* (Mayle 1993); a coffee-table photographic account of the Provençal landscape; a watercolour version of *A Year in Provence*; a perfume; and red and white Côte du Lubéron wines labelled 'A year in Provence', stamped with Mayle's signature and marketed by the English supermarket chain Sainsbury (Aldridge 1995).

A Year in Provence and *Toujours Provence* detail Mayle's 'escape' to rural France. He writes that he chose to give up the hectic life of an advertising executive to move into semi-retirement in a farmhouse in the Lubéron district of Provence to pursue his new vocation as a novelist. Mayle is accompanied by his wife, about whom the reader learns very little (not even her name) except for his occasional references to 'my wife' or '*Madame*' and casual remarks about her love of dogs, her appetite and her ability to come up with an idea 'that only a woman could have had' (*Year*: 185). In contrast, Annie Mayle has a prominent role in the TV adaptation.

The first book, *A Year in Provence*, is organized into monthly chapters which catalogue the Mayles' arrival in January, and their subsequent initiation into local customs and cuisine, their attempts to deal with French bureaucracy and local workmen, and their growing friendship with their neighbours. The book is sold as 'an account that attempts to answer the question: What is it *really* like to live in Provence?' (*Year*: back cover). The follow-up volume, *Toujours Provence*, presents a more thematic and detailed account of different aspects of Provençal culture, including the intrigue of buying truffles, Mayle's quest for 'the singing toads of Pantaléon', the craziness of Provence in August and his and his wife's 'going native'. In contrast to the dull, drab – and invariably wet – descriptions of England, the books vividly portray the sights, sounds, smells and most importantly the tastes of rural Provence.

The much less successful book, *Hotel Pastis*, is a fictional account of an executive's move to Provence to escape the high-stress life of London and to open a hotel in the Lubéron. The executive also escapes a society ex-wife to find new love in Provence. In addition to the rather convoluted plot twists that account for a botched bank robbery and kidnapping, the book provides details that I can only assume throw more light on Mayle's own move. In this case, the protagonist defends tourism to a local English sceptic, explaining the benefits that the industry can provide as long as the mistakes of the past mean that some lesson has been learned (*Hotel*: 293). This, I expect, is in some way a reaction to those critics of *Year* and *Toujours* who claimed that the two books had reinforced the negative effects of tourism on Provence – having 'ruined life in the Lubéron' (Paul Eddy in *The Times* quoted in Lefort 1993, 14) – as a result of their popularity.

Reviews of the books have been mixed, with greater criticism of the later works. On the one hand, a *Times Literary Supplement* reviewer had no doubts over the realism of Mayle's work. He claimed that Mayle 'appears to understand and be naturally sympathetic to the points of view of all manner of local people' (Winterbotham 1989, 844). However, others have not been nearly so generous, lambasting Mayle for his 'limp humour, the self-congratulatory tone', but more importantly, for his insensitivity to the wonders of Provence: 'the almost complete absence of evocation of place, history, the astounding light' (Thorpe 1991, 21). Mayle's works have become a symbol for all that is wrong with middle-brow culture in the eyes of English élites and intellectuals.[1] They argue that 'Peter Mayle induced romanticism' (Fraser 1996, 25) has allowed an 'anything goes' attitude to all things French.

The television adaptation is based upon a mixture of *Year* and *Toujours*, with a great deal of additional material that appears in neither. This account is not nearly so descriptive as Mayle's originals, instead being focused around a series of laboured stories about local goings-on and intrigues. Significantly, the books' idealization of the French countryside and the humour found in local idiosyncrasies and the irritable visitors from England are mostly replaced by conflict: Mayle's reclusive neighbour Massot (Rivière in the TV version) becomes a repulsive, sinister and almost fearful figure; there is a running battle with various tradesmen as Mayle becomes increasingly irritated with their laxity; and English Tony is actually thrown out of Mayle's house (whereas in the book, his inability to deal with the French property system has him removed from the tale with a much more satisfying sense of just desserts). In contrast to the books, which present Mayle as a fluent Francophile, intent on 'fitting in' and doing everything the local way and at local speed, the TV version has him attempting to communicate with locals in a ridiculous 'franglais' and becoming increasingly irritated at their way of doing things, especially the time they take to do them. The TV version was a failure, consigned to infamy as one of the BBC's more recent disasters.

It seems that it is the non-conflictual rural idyll that was behind the books' popularity. Mayle achieved this effect via two textual devices: an 'anthropological' mode of writing and the construction of an image of community which resonates with the *Gemeinschaft* of traditional belief systems in comparison with the mediated *Gesellschaft* of Mayle's previous way of life.

LITERARY DEVICES

Western travellers have tended to adopt a colonialist style of writing which assumes the superiority of the traveller's cultural and moral values and which leads to this figure taking possession of what he [*sic*] sees in a voyeuristic gaze (Pratt 1992; Blunt and Rose 1994). Even when sympathetic towards the people being visited, this colonial rhetoric positions the indigenous people as childlike or lacking in reason. Variations on this general theme have characterized travel within Europe. 'The Grand Tour' took élites on cultural pilgrimages to the great sights/sites/cites[2] of 'civilization'.

Travel in the form of (semi-)permanent migration to rural areas is obviously based upon motivations and couched in expectations that differ from travel outside Europe. However, the parallel with the colonial traveller is often maintained, especially when the destination is the countryside. Despite the modern-industrial realities of much of the countryside in Europe, it is predominantly represented in the English media as immediate, natural and timeless in contrast to the modernity and rationality of metropolitan life. Within English thought, rural dwellers have occupied a position similar to individuals formally colonized by metropolitan powers.

Lady Fortescue's account of her move to Provence in the 1930s clearly illustrates this colonial mode of writing the rural: she acts as the figure to whom others look up, and speaks of the French peasants as if they were children. She describes a scene of childlike wonder when the otherwise gruff removal men unwrap her possessions, which have just arrived from England: 'They had naturally never seen anything like it before. They stroked the little beasts with grimy forefingers and noticed their gleaming eyes' (Fortescue 1935, 27). Obviously Fortescue had not moved to France as part of the institutions of colonial exploration, exploitation or rule in any direct sense.[3] However, her attitude to local peasants is clearly not dissimilar to English attitudes towards colonized people around the empire.

Mayle's country folk are also simple peasants leading uncomplicated, rather idealized lives, compared to the sophistication of the city dwellers of London, Paris and California. What is different about Mayle's delineation of city and country though is his valorization of the rural ways and his ridiculing of over-cultured urbanites. This is achieved through his adoption of an anthropological rhetoric that imbues all of his work.

Mayle the anthropologist[4]

As the title of *A Year in Provence* might suggest, Mayle adopts an anthropological mode of writing. Local words and meanings are revealed to the readers as being key to understanding how things are. This mode presents Mayle as deferring to local knowledge, in contrast to earlier accounts, such as Lady Fortescue's, whose more colonial modes of reporting can be seen to impose an English sensibility upon that of the locals.

Mayle positions himself as an informed insider reporting his findings to his readers. In his texts he foregrounds his role as translator of Provence for his English audience. He quite consciously positions himself as a hybrid lying somewhere between England and Provence, understanding and valuing both: 'It was the kind of meal that the French take for granted and tourists remember for years. For us, *being somewhere between the two*, it was another happy discovery to add to our list' (*Year*, 181; emphasis added).

Yet Mayle should not be seen as a post-colonial figure. His humorous reporting of events may undermine the authority of modern urbanites, but the rural population does not remain unscathed. Despite the appearance of deference, and the cover blurb's proclamations of his fondness for Provence, overall Mayle's tone is rather paternalistic. His valorization of local ways is somewhat deceptive. Although he ridicules city people and their ways, he also pokes fun at the residents of the Lubéron to the extent that he has been accused of being patronizing towards them (for example, see Thorpe 1991). This would appear inconsistent with his own claims that he himself has become a 'bumpkin' (*Year*, 128) as a result of 'going native' (*Toujours*, 202). Yet, he is apparently conscious of his readership. He constantly reminds the reader that he is an outsider, both directly and through his construction of *the* Provençal character. His narrative comes from a position that takes local ways seriously but which is also informed by an urban sophistication that allows him to make jokes at their expense that he know his (mostly urban) readership will understand. Thus, although he identifies himself with the natives, this is to ensure he is not confused with the pretentious, insensitive 'rural tourists' from cities. He is always clear to ensure that the reader knows his origins, knows that he is a sophisticated urbanite, that there is no way that Mayle could possibly be as straightforward and basic as the rural folk he describes. He positions himself as an informed outsider reporting his privileged information to other sophisticated urbanites.

When reporting Provençals' words in English, Mayle is able to make his characters intelligible to his audience while still maintaining an air of difference by offering excessively direct or arcanely idiosyncratic translations. He also includes many French words in the text, words that remain untranslated for the most part but whose meaning is obvious in context, or from school or 'menu' French. The French words interrupt the neat flow of Mayle's narrative (if for no other reason than their italicization), reminding the reader that the account is of a place other than their own.[5]

Perhaps more significantly, Mayle writes out the Provençal dialect. Here he reminds the reader that he is talking about a particular place: he is indicating even to those who speak textbook French that he is recounting a specific place, a place that cannot be understood from reading, but must be experienced – must be heard in the original Provençal pronunciation – in order to understand it. But even this is insufficient. Mayle's Provence is a total, embodied experience: communication is in excess of the efficiencies of the spoken word. Instead Mayle's Provençals take great delight in their performance:

The words alone do not do justice to the occasion, which is decorated with shrugs and sighs and thoughtful pauses that can stretch to two or three minutes if the sun is shining and there is nothing pressing to do.

(*Toujours*, 212)

Indeed, many things can be understood only as a result of their role in the rituals of everyday Provençal life. Rituals of consumption of food and drink provide numerous examples of this localization of experience. For example, Mayle describes the local drink, *pastis*:

> For me, the most powerful ingredient in *pastis* is not aniseed or alcohol, but *ambiance*, and that dictates how and where it should be drunk. I cannot imagine drinking it in a hurry. I cannot imagine drinking it in a pub in Fulham, a bar in New York, or anywhere that requires its customers to wear socks. It wouldn't taste the same. There has to be heat and sunlight and the illusion that the clock has stopped. I have to be in Provence.
>
> (*Toujours*, 157)

Thus in contrast to the detached gaze of the tourists that he sees in the Lubéron, 'drift[ing] through the streets of Lacoste and Ménerbes and Bonnieux in a sight-seer's trance, looking at the people of the village as if they too were quaint rustic monuments' (*Year*, 125), Mayle's account of Provence is very much an embodied one. He constructs his place from the combination of the sounds, tastes, feels and experiences of Provence in addition to the sights. The locating of Provençal dialect and the affirming or contradictory effects of hand gestures reinforces this importance of *being there*, of being a part of the community, not just an observer of it.

Mayle constructs Provence not through the production of spectacles of difference, nor by the monuments and landmarks that guidebooks mark as making up the uniqueness of the region, but rather through his interest in the details of its daily workings. The book describes the everyday life of the locals and the adaptations that the Mayles make in order to fit in with this new way of life. English friends who were surprised at the Mayles' apparent lack of activity wondered whether they got bored with everyday life. Mayle responded, 'We didn't. We never had time. We found the everyday curiosities of French rural life amusing and interesting' (*Year*, 73). Mayle takes great effort to describe the colour of the details of Provençal life, indicating that in no way could his life be described as bland or boring. This is most emphatically demonstrated in his description of the reactions of an English guest, Susan, to her stay in Provence:

> She was allergic to the South. . . . It's not uncommon. Provence is such a shock to the Northern system; everything is full-blooded. . . . There is nothing bland about Provence, and it can poleaxe people as it had poleaxed Susan.
>
> (*Year*, 106)

Mayle's emphasis upon the details and colour of the everyday has four rhetorical effects. First, it infuses the texts with traces of the anthropological modes of writing with which he clearly wishes to have his work resonate. With an anthropologist's almost obsessive interest in the mundane everyday rituals of their subject community Mayle constructs Provence out of myriad little differences. Secondly, the richness of detail provides his books with a 'reality effect' which works towards making his account seem believable as a result of the great detail he provides. Indeed, so well has the realism been conveyed in Mayle's writing that one reviewer suggested that 'the accuracy of the author's descriptive observation make it advised reading for anyone planning to move to Provence' (Winterbotham 1989, 844). Thirdly, in its use of all senses, the detail provides the richness of the embodied experience of being there, contrasting Mayle's experience to the passive voyeurism of mass tourism. Finally, in its provision of the immediacy of the Provençal way of life it imbues the community described with a sense of *Gemeinschaft*, evoking the community identity which, I will argue, is central to the draw of the experience described in *Year* and *Toujours*.

Provence as pastoral England

If 'the past is a foreign country' (Lowenthal 1985) then in this case a foreign country is the past: Provence in Mayle's accounting is an English past. It is an idealized England organized around a productive, but non-industrialized countryside from which emerges a landscape of consumption for its viewers. Furthermore, as if to reinforce the imagined value of this particular community, it can provide a rural idyll of England past in which the sun always shines.

In a more economistic vein, Buller and Hoggart (1995, 2) have suggested that contemporary migration to rural areas such as the Lubéron can be described as 'post-Fordist,' determined not by work-related moves, but by a stress on individuality and consumer preference. An economic rationale for moving to the French rather than the English countryside emerged in the late 1980s as a result of rapidly rising English house prices, particularly in the south-east of the country. Added to this is the fact that the industrialization of most lowland English countryside has meant that affordable *pastoral* rural dwellings were hard to come by. The French countryside has been altered by industrialization to a much lesser degree (Buller and Hoggart 1995). As it has also been less densely populated, rural property was more readily available and relatively inexpensive in the late 1980s (Ibid., 33). 'Go south to the sun to beat the recession', the English tabloid press clamoured (Gates 1992, 40), confirming this trend.

'The countryside' has great cultural significance for the English middle classes, whose cultural 'habitus' (Bourdieu 1984) is formed around an anti-urbanism and anti-industrialism (Wiener 1981; Williams 1973) which holds pastoral England to be the threatened heart of national identity. Many English migrants to rural France explained their move to Buller and Hoggart (1995, 4) as motivated by a desire to recover an 'aura of genuine rurality' that many think has been lost from lowland Britain.

Mayle describes a lifestyle that is simpler and less developed and complicated than that which he (and his assumed audience) have experienced in England (and other modern, urban societies). His French countryside is run not by agri-businessmen engaged in industrial farming but by peasants whose ties to the land go back for generations, who hate to see waste, who do things by hand, who wear traditional attire and who have good, honest values. For example, Provençal identity is written as simple and honest in contrast to the preenings and sophisticated fashions of city dwellers. This is written in a number of ways, one of which is through a language of 'healthy' sexuality. Provençal masculinity is presented as being rugged and earthy in comparison to the effeminate men from cities. When Tony arrives from London, he 'stood by the swimming pool, fending off [the Mayles' dogs'] affections with a *handbag of masculine design* . . . he and his handbag joined us for breakfast' (*Year*, 59, emphasis added). His masculinity is further ridiculed in Mayle's subsequent description of his pastel-coloured clothes, his 'carefully touseled hair', his gold lighter and watch; Mayle speculates that 'gold medallions nestled in his chest hair' (*Year*, 59). In contrast, the men of Provence are confident in their sexuality and shock outsiders with their acts of affection. Mayle reports his London lawyer friend's reaction to Provençal greetings:

> 'Look over there,' he said, as a car stopped in the middle of the street while the driver got out to embrace an acquaintance, 'they're always mauling each other. See that? Men *kissing*. Damned unhealthy, if you ask me.' He snorted into his beer, his sense of propriety outraged by such deviant behaviour, so alien to the respectable Anglo-Saxon.
>
> (*Year*, 94)

However, Mayle's vision of Provence is not purely a Rousseau vision of lost innocence and an outright rejection of progress. Mayle is not a simple romantic.

Year and *Toujours* clearly valorize the Provençal lifestyle over modern urban-ized ones, but Mayle's romantic escape was only possible in the first place because of the world he sought to escape. First, his position in high capitalism was a precondition to his ability to establish a life in Provence. To afford to live in semi-retirement in another country in the hope that his new career of fictional writing would sustain him required a sizable initial capital investment. Others who have pursued this ideal have found it impossible to continue because of the cost: the average stay of British immigrants in rural France is around two or three years (*A French Affair*, Channel 4, 28 March 1994).

Secondly, Mayle's publications are caught up in a central process of moderniza-tion: tourism is itself a key facet of modern life – and its existence was necessary for the success of *Year* and *Toujours*. These two books are very much a part of modern life, not some kind of escape from it. Mayle is clearly not unaware of the importance of international cultural productions in the construction of 'authen-tic' experiences. In *Toujours* he describes watching a truffling pig and imagines the scene set to the theme music of *Jean de Florette*, the very successful French film about traditional Provençal life. This rhetorical move gives his second book added intertextual legitimacy of an internationally recognized 'local' cultural production.

Mayle accepts the necessity for progress, unsettling its excesses with humour rather than morality, with a clear belief that Provençal values will survive change, as they have in the past. Instead what Mayle apparently objects to is the dominance of the political economy of the sign characteristic of Thatcherite England, especially the south-east, and especially in an industry such as advertising.

AN UNMEDIATED PLACE

> The effect of the weather on the inhabitants of Provence is immediate and obvious.
>
> (*Year*, 8)

That Mayle is not simply constructing an opposition between the traditional and modern is clearly illustrated in a number of anecdotes, but most clearly in the long narration of the Mayles' struggle to have central heating installed in their farmhouse. And, despite his stereotyping of peasants and their values, Mayle does not present them as being entirely untouched by progress – indeed he recognizes their eagerness to adopt certain aspects of modern life, especially the labour-saving ones. Even 'Grandfather André', a figure who might be seen to represent the voice of Provence's past within Mayle's narrative, insists on leaving the (manual) vine planting early 'so that he could watch *Santa Barbara*, his favourite television soap opera' (*Year*, 44). Mayle describes the plumber's enthusiasm for the most modern equipment and gadgetry, and his description of his neighbour's fox recipe gives a further sense of the Provençals' ironic interaction with progress: 'In the old days, this was eaten with bread and boiled potatoes, but now, thanks to progress and the invention of the deep-fat fryer, one could enjoy it with *pommes frites*' (*Year*, 17). So the peasants are not simple people, unaware of the march of time after all. Indeed it is Parisians and other 'sophisticated' city types that Mayle holds up to ridicule for see(k)ing authenticity throughout Provence. In a Parisienne's soirée, 'there were some refined squeals of appreciation at the deliciously primitive setting – *un vrai dîner sauvage* – even though it was only marginally more primitive than a garden in Beverly Hills or Kensington' (*Year*, 128). Mayle offers a picture of a place not devoid of change or progress but whose syncretic relationship with change is on the whole neither conflictual nor damaging but humorous (such as the availability of *pommes frites*, or the plumber's obsession with hi-tech toilets). This change is superficial. In the end then, Mayle presents the Provençals' interaction with the agents or symbols of change as one which allows them to exhibit their unique identity rather than seeing it fall away under the pressures of global culture.

Instead of simply seeking some form of 'authentic' peasant life, Mayle seems to have been in search of a way of life in which, in sociological terminology, social integration dominated over systems integration: he was in search of *Gemeinschaft*.

Similarly, a young English couple who moved to rural France in the late 1980s explained the motivation for their uprooting as a reaction to that decade's 'have everything' culture, a culture which they saw as being 'very false and unrealistic' (*A French Affair*, Channel 4). This search appears to be central to much contemporary travel writing and tourism and is articulated perfectly by an obviously wealthy woman traveller in a tent in Mongolia speaking on BBC's 1996 series, *The Tourist* (7 January 1996):

> We live in a totally symbolic world. Everything we do is a symbol. . . . We hand over bits of metal or paper and come back with a plastic box with M-I-L-K on it and nothing has any reality at all. We don't make anything with our own hands anymore. We just shuffle papers and push buttons and I think that as those things are unreal, they have no value. They don't feed you: they don't feed your intelligence, they don't feed your soul, they don't feed any human necessities.

It would be hard to find a better example of the invasion of the sign than in the world of advertising that Mayle 'escaped'. For him, and the woman quoted above, life had become too mediated: in more abstract terms, their life had been 'invaded' by systems integration, their experiences given meaning only through deferral to extra-local systems of value, distinction and meaning. Provence as described by Mayle, on the other hand, provides a life structured around social integration, of face-to-face contact, of kinship and social solidarity:

> You can live for years in an apartment in London or New York and barely speak to the people who live six inches away from you on the other side of a wall. In the country, separated from the next house though you may be by hundreds of yards, your neighbours are part of your life, and you are part of theirs.
>
> (*Year*, 4)

Mayle contrasts the dominance of 'natural' use values in Provence to signs of transitory (fashionable) meaning in urban culture. His description of his reaction to coming upon the smell of a fire illustrates this well. He claims that '[wood smoke] is one of the most primitive smells in life, and consequently extinct in most cities, where fire regulations and interior decorators have combined to turn fire-places into blocked-up holes or self-consciously lit "architectural features"' (*Year*, 10). Mayle is suggesting that what has value in Provence for what it is and what it can do has value in cities only as a result of what it can represent. Similarly he contrasts the Provençal markets displaying freshly picked local vegetables, local produce and animal carcasses to city supermarkets where 'flesh was hygienically distanced from any resemblance to living creatures. A shrink-wrapped pork chop has a sanitised, abstract appearance that has nothing whatever to do with the warm, mucky bulk of a pig' (*Year*, 9).

To return to our Mongolian traveller's rhetoric, Provence offers Mayle a wealth of *imm*ediate consumption: as the Mayles exist as non-agricultural workers living in the countryside, the rural landscape offers itself for their visual pleasure, but more importantly, the bountiful nature of locally produced food and wine offers them enough to engorge their 'human necessities'.[6] But this is a form of consumption without the guilt of accumulation. I mean this in two different ways. First, and most literally, the Mayles enjoy the wonders of the Provençal cuisine and yet also manage to lose weight in the course of their year, surely an ideal in a time when individuals are bombarded with images of both consumption and the necessity of maintaining the body beautiful.[7] This is in contrast to the anorexic diet of Parisians, Londoners, New Yorkers and – worst of all – Californians. Mayle presents these diets as being a very unnatural relationship to consumption: they are 'diet-crazed' (*Year*, preface).

But secondly, Mayle describes a place in which there is no unequal accumulation, none of the economic exploitation characteristic of (urban) late capitalistic social relations. The arrival of Mayle and his wife proves to be only beneficial to locals in an economic sense and does not disturb the socio-economic 'equilibrium' already in place. Their relations are forged around a concept of reciprocity in that they are 'engaged in mutually profitable relations with artisans' (Aldridge 1995, 425). Mary-Louise Pratt has described this rhetoric as one central to Western capitalism's self-image, especially as written in European travel writing (Pratt 1992). The concept that they come as friends rather than individuals mediated into society by economic relationships alone is very important to Mayle – as it has been to the literature of Western travel. What this rhetorical strategy achieves is a resonation with the utopia of capitalism in which exchange is mutually beneficial and where there are no distinct or uncomfortable class divisions. This stands in stark contrast to the competition and commercial over-development of both the Thatcherite society that he 'escaped' and the economic relations foregrounded in mass tourism.

Over the course of the two books Mayle describes his developing role as part of the Lubéron community: he and his wife are known locally as *les Anglais* (indeed they are marked on a map as such in the village shop), but this is not intended simply to mark them as outsiders: they are the Lubéron's English. Mayle attributes this real community membership, first to their permanent residence and this demonstration of true commitment to the place, and secondly to their acceptance of Provençal ways. They are not merely passive voyeurs of Provence, but, as already mentioned, they are integrated socially. It is implicit in Mayle's understanding of things that he and his wife are members of the community, enforcing established relationships rather than distorting them. It is tourists who exist in deformed economic relationships when visiting a place. Mayle reports his peasant neighbour's complaint that '"Germans with tents don't buy anything except bread"' (*Year*, 47) The Germans' relationship with Provence is mediated purely by economic relations. Locals will do things for tourists only when payment is involved, whereas the Mayles' relationship with Provence is mediated by friendship and support – locals are keen to show them things and explain ways. Even when money is involved, as with the Mayles' building works, the workers present a gift to welcome them to the community. This romanticization of social integration almost harks back to the importance of the exchange of gifts in what Sahlins (1974, 186) has famously termed 'Stone-Age economics': 'If friends make gifts, gifts make friends.'

Mayle pours derision on the Côte d'Azur in summer where cynical locals attempt to extract as much money as possible from tourists. Here the friendship which is offered to visitors is all too staged in the pursuit of one thing – money:

> A hostile cupidity hangs in the air, as noticeable as the smell of Ambre Solaire and garlic. Strangers are automatically classified as tourists and treated like nuisances, inspected with unfriendly eyes and tolerated for cash. According to the map, this was still Provence. It wasn't the Provence I knew.
>
> (*Year*, 109)

And with these new values, new dangers emerge: residents had to construct elaborate security measures to protect their homes, losing the cultural and social connectedness that Mayle so treasures. Mayle describes a kind of fortress Provence on the coast in terms not entirely dissimilar to Mike Davis's *City of Quartz*, Los Angeles. The trust and community dependence characteristic of the Lubéron are missing from this version of Provence.

As a result, Mayle's greatest derision is reserved for fashionable people in general, and Parisians in particular. Parisians seek to transform Provence into their own quaint photo shoots and soirées. In *Toujours Provence*, Mayle fears the worst: 'Provence has been "discovered" yet again – and not only Provence in general, but the towns and villages where we shop for food and rummage through the markets. Fashion has descended upon us' (*Toujours*, 114). This 'invasion' is not new or unexpected – its results can be predicted all too well:

> It has happened before, in many other parts of the world. People are attracted to an area because of its beauty and its promise of peace, and then they transform it into a high-rent suburb complete with cocktail parties, burglar alarm systems, four-wheel-drive recreational vehicles and other essential trappings of *la vie rustique*.
>
> (*Toujours*, 125)

Certainly Mayle has historical precedent for his fears. Much of the Mediterranean has been changed as a result of the influx of visitors and their expectations of what that area should be like. Indeed the reminders of the effects of tourism are close by in neighbouring Côte d'Azur, an area which has had links with British travellers since the turn of the century, and which became established as a place to get away to long before other Mediterranean tourist developments. Now, the 'Côte d'Azur is no longer a place for escape' (Blume 1992, 172), but a place that looks elsewhere for its standards and values – a place of oil-free salads, high-rise buildings and fear of crime – all those things most repulsive to the Provençal society written by Mayle.

What threatens the Lubéron for Mayle is not progress *per se* (which *could* facilitate a sensitive development of tourism) but fashion – the intervention of the symbolic into the real working of things so that value is only obtainable through deferral of meaning to style gods in Paris or the pages of *Vogue* magazine. In the society constructed by Mayle, it is a sense of belonging, of shared meanings and values derived from local issues and the obligations of friendship which construct meaning, not some distant symbolic order. Fashion and popularity will bring to Provence what ruined the coast: fashion statements, instant communication via fax machines, the latest fad diets . . .

But on the other hand, Mayle is not so gloomy for the future of Provence. First, the problem of tourism that he narrates is a temporal one, the effect of the 'natives of August'. Mayle realizes that 'we live in the same house, but in two different places' (*Toujours*, 175). At a soirée he remarks to his hostess that the party 'didn't seem at all like Provence, and she shrugged. "It's August" ' (*Year*, 128). There is chaos and overcrowding and commercialism during July and August, but come September, claims Mayle, things return to normal. It is accepted that life changes for a couple of months a year. But the tourist industry provides money, employment and, not least to Mayle and his Provençals, a stock of humorous stories for the remainder of the year.

Tourists also provide locals with a scapegoat for all problems. Mayle lists the national stereotypes produced by the Provençals to explain all problems: littering German tourists, bad Belgian drivers, the English and the gastric complaints and so on. Yet Mayle suggests that it may not always be the tourists at fault. He describes his own experience of campers:

> They were mostly Germans, but not the indiscriminate rubbish tippers that [Massot] complained about. These Germans left no trace; everything was bundled into giant backpacks before they shuffled off. . . . In my short experience of litter in the Lubéron, the French themselves were the most likely offenders, but no Frenchman would accept that. At any time of year, but particularly the summer, it was well known that foreigners of one stripe or another were responsible for causing most of the problems in life.
>
> (*Year*, 111)

Ultimately, Mayle believes that Provençals are not so fickle as to accept changes without thinking through their consequences. Despite the popularity of Provence as a tourist destination and the changes that this has brought, Mayle writes of people in the Lubéron who have not readily accepted the dominance of 'the sign'. This is narrated in his account of the arrival of Tony from London. Tony explained to Mayle that

> money was an international language and he didn't anticipate having any difficulties. Unfortunately, when the bill arrived he discovered that neither

his gold American Express card nor the wad of traveller's cheques that he
hadn't had time to change were any interest to the restaurant's proprietor.

(*Year*, 63)

This tale can be seen as a synecdoche representing Provence's refusal to submit
to globalization more generally. Furthermore, Mayle acknowledges that 'traditional' and 'modern' systems could probably coexist. In fact they do already, as
is evidenced by the meaning of 'Provence'. According to *Year* and *Toujours*, for
the Mayles and the locals Provence is based on community and the seasonality of
life, whereas for tourists the place is based upon deferral to an internationally
agreed inventory of 'Provence':

> With so many distractions on our doorstep, we were neglecting the more
> famous parts of Provence, or so we were told by our friends in London. In
> the knowledgeable and irritating manner of seasoned armchair travellers,
> they kept pointing out how conveniently placed we were for Nîmes and Arles
> and Avignon, for the flamingoes of the Camargue and the *bouillabaisse* of
> Marseille. They seemed surprised and mildly disappointed when we admitted
> that we stayed close to home, not believing our excuses that we could never
> find the time to go anywhere, never felt a compulsion to go church-crawling
> or monument-spotting, didn't want to be tourists.
>
> (*Year*, 96–7)

Although his romantic anthropological perspective can produce an anxiety that
cosmopolitanism might threaten the distinction and difference of each place,
Mayle acknowledges that the more realistic Provençals are quite happy with this
situation; for many of them, cultivating tourism is easier than and preferable to
the traditional peasant farming way of life. He explained the benefits in *Hotel
Pastis*:

> . . . the people at the Brassière saw the hotel as a source of diversion and
> possible profit. Their properties would increase in value; there would be
> more jobs; perhaps their children wouldn't have to leave the village to find
> work – for them, tourism was attractive. The postcard version of the peasant's
> life, picturesque and sunlit, was a long way from the grinding reality of disap-
> pointing crops, aching backs, and bank loans.
>
> (*Hotel*, 216)

Despite grumblings about the inconveniences caused by the influx of people into
the area, locals 'admitted, with much nodding and sighing, that tourists brought
money into the region' (*Year*, 125).

At times Mayle goes even further to suggest that traditional Provence will
outlive fashionable changes, superficially changed with more conveniences but
essentially the same. He sees Provençal stability compared to faddish vogues. For
example, in Aix, the Mayles would window shop 'to see what new nonsense is in

the windows of the boutiques which are crammed, chic by jowl, next to *older and less transient* establishments' (*Year*, 100; emphasis added). In other places, as in the long quotation from *Hotel Pastis* above, he appears to think that tourism might actually reinforce the traditional way of life by making small rural communities financially viable.

The real victims of more tourists coming to Provence are not the locals but the incomers like Mayle, those who came in search of the 'positional good' of peace and quiet and, most importantly, cultural difference. It is the foreign immigrants who want to keep other foreigners out of their idyllic landscape. For example, Buller and Hoggart (1995, 118) report that one British migrant to the Dordogne region opined that the 'French should impose an embargo on people from foreign countries in areas of concentration'.

It is possible to view Mayle's accounts as playing a therapeutic role within wider narratives of modernization and Westernization, but particularly in tourism. Although he recognizes the effects of tourism and modernism and change, he does not see them as bludgeoning difference from communities, except when managed badly. Each place will react and respond differently so that variation will remain. In contrast to the post-colonial anxiety of many travel writers, who report with growing dismay of the loss of authentic culture and, they assume, difference, Mayle writes Provence as more resilient in its maintenance of identity. In a sense, he allows travellers to have their cake and eat it: it is all right to be a tourist, you are welcomed for the money you bring in, and, if you leave your urban ideals in the city, there will not be too much damage done.

CONCLUSION: 'PROVENCE' – JUST A FICTIONAL LANDSCAPE?

Mayle has been accused of simply producing English myths for English audiences (Aldridge 1995, 417). I think that this is true, but should not be overstated. Edward Said has suggested that Orientalism 'has less to do with the Orient than it does with "our" world' (Said 1979, 78). This does not mean that Orientalism had *nothing* to do with the territory encompassed by the concept of 'Orient', and so it is with Mayle's Provence. His romantic anthropological style re-creates a bountiful rural landscape apparently no longer in existence in England: a society based around community values, not just symbolic exchange. But although Provence is in excess of Mayle's accounts of it, it would be misleading to suggest that the books have nothing to do with the place.

Concern over the 'authenticity' of Mayle's representations of rural France has led some commentators to accuse Mayle of influencing the development of Provence, by enticing more and more English tourists with his idyllic vision of traditional English values embedded within Provençal life. There are fears that the Lubéron will become overdeveloped and 'corrupted' like its Côte d'Azur neighbour. Mayle's anti-fashion has made Provence fashionable for British travellers; he is mentioned in the majority of British guidebooks to the region – even located on some maps. It has been suggested then that he has helped to rewrite the land-scape, that the books do not remain in the circulation of knowledge and images but have become actualized in the landscape of Provence itself, so 'ruining' it for others. However, I do not believe that this argument should be taken too far.

First, in their study of British migrants to France between 1980 and 1991, Buller and Hoggart (1995) found no evidence of significant impact of Mayle's work (although admittedly this study only covered the years immediately follow-ing publication of Mayle's books). Over the course of this period, although the number of Britons in the Vaucluse area increased slightly, there was a decline in the percentage of British migrants to France coming to this area. In other words, Provence was not growing as a destination for British migrants to the extent that other areas were. Despite fears of foreign overcrowding by British immigrants, Buller and Hoggart (1995) concluded that in all areas of their study the percent-age of migrants was small, rural landscapes remained open and population levels in the countryside were low. Furthermore, they argued that most areas had ben-efited from British migration, with improvements being made to a housing stock that had lain unoccupied, being of no use to local populations and no interest to French urban dwellers (Buller and Hoggart 1995, 99). Most of the people they interviewed seemed to feel, like the Mayles, that they had been made welcome in the communities they had joined because they made the effort to integrate into local communities and because, as one woman claimed, representative of many, 'They were so pleased that I was not a Parisienne – they are the anathema of the French country people' (quoted in Buller and Hoggart 1995, 106). This view echoes a very clever rhetorical strategy of *Year* and *Toujours*. Mayle has managed to write himself as an Englishman into the Provençal society – a feat he claims that Parisians cannot achieve. In narrating his interaction with the Provençals, he erases his foreignness because of the values he shares with them, and the effort he puts into 'fitting-in' to the community. In doing so, he challenges the homo-geneity of national culture and as a result displaces concerns about the negative impact his 'foreign' presence might have on the place.

Secondly, Mayle cannot be accused of initiating a rush to Provence because Provence has been a favoured destination for British travellers and tourists for decades, stemming from the popularization of the Côte d'Azur by English upper classes in the later nineteenth century. Even discounting the Côte d'Azur, this is an area that has witnessed large-scale migration so that 'only very few Provençals today can claim a local origin that dates back more than one or two generations" (Lefort 1993, 14). The authenticity that Mayle seeks to present is an invented tradition already.

In his book *Late Imperial Romance*, McClure discusses a concept closely related to the paradox of tourism mentioned in the introduction. Drawing on Frederic Jameson, McClure (1994, 9) defines romance as being 'designated to satisfy collective longings for adventure and Otherness, "magic and providential mystery"'. But that which seeks to relate the excitement of these romantic places to others has ended up destroying them by making them known and so removing their mystery. McClure continues, 'when it became impossible to ignore the prospect of global modernism, the eradication of the last elsewhere, the writers of imperial romance began to become uneasy' (Ibid., 11). In writing his Provence, Mayle at first offers a new romantic space of community life and easy living, in distinction from the fashionable images that had dominated. He offers an alternative to mass tourism: he contrasts his insider status with tourists who remain on the outside, linked to the place being visited only by economic relations. And yet Mayle's therapy has limited effect. His books and their spin-off products themselves became fashion: his rebellion against the Anglicized, Americanized, Parisian-ized landscape of Provence became fashionable itself. Mayle's Provence entered the commodity circuits of image production – but became overproduced. Mayle's Provence became mass produced and as a result could not provide the distinction required by bourgeois habitus. Bourdieu has suggested that the bourgeois prefer 'organized, signposted cultivated nature [and culture]' (quoted in Urry 1990, 89). To enjoy distinction from mass tourism – an enjoyment of travel organized around the romantic gaze of the lone individual – requires knowledge of the place being visited, provided in many cases by travel books. Problems arise when a travel book becomes popular so that its knowledge becomes codified in the popular consciousness – and at times in the landscape itself – and becomes a new set of 'signposts' for the bourgeois traveller. It appears that this has occurred with the *Year in Provence* phenomenon. It should perhaps come as no surprise that Lady Fortescue's *Perfume from Provence* has become the middle class's new talisman of rural France. This is not simply because it is somehow regarded as more authentic. It is more different, being as it is removed in time as well as in space. The aristocratic writing of Provence constructs a true positional good that cannot be obtained and so cannot become passé like Mayle's depictions. Now with the commodification of travel and travel writing it would seem that a foreign country really can only be the past.

As if to confirm this, the place that Mayle so lovingly depicted has now changed for him so much that he has felt compelled to move on himself – in order to achieve the type of lifestyle of his year in Provence, he has moved to the north-east USA.[8]

NOTES

1 Indeed the greatest critical insult that a book reviewer on BBC2's arts programme *Late Review* could level at Julian Barnes' collection of short stories about Anglo-French relations, *Cross-Channel*, was to state sarcastically (and as it transpired, hyperbolically) she had learned more about France from Peter Mayle.

2 I adopt this term from Ó Tuathail (1994, 535), who has applied it to the discursive practices of statecraft, but I believe that it is also appropriate to the practices of travel and tourism. The three interdependent terms of sight, site and cite are key to the experience of travel: the privileging of the visual, the location of the traveller in key historical locations, and the intertextuality of places on the tourist trail.

3 Furthermore, her role as a female narrator would suggest a different relationship to power than if *Perfume from Provence* had been written by a man (see Pratt 1992; Blunt and Rose 1994).

4 Aldridge (1995, 415) suggests that Mayle adopts a 'self-consciously "anthropo-logical mode"' in order to differentiate himself from travellers writing in a colonial mode.

5 What can be presumed to be Mayle's intended effect of his use of French words in the text is not necessarily how readers will interpret this strategy. It thus comes as no surprise that not all commentators have accepted Mayle's use of French words in the text positively. A reviewer of *Toujours Provence* considered that this textual strategy reinforced his caricatures of French peasants in that their speech 'is peppered with native *bon mots* perilously close to that of *Allo, Allo*' [a British sitcom comprised of stereotypical French characters whose use of exaggerated French terms was the root of much of the otherwise slapstick humour] (Thorpe 1991, 21). Indeed I would suggest that the TV version does indeed stray into *Allo, Allo* territory, relying as it does on the basic 'smutty' humour of French (mis)pronunciations of English words.

6 So central is this aspect of *Year* that it makes up the most important, and most clever, part of Martin's (1992) satirical version. In his account, set on the planet Provender, the whole environment facing the immigrants is available for consumption . . . as indeed are its residents. To their peril, the author and his wife discover at the very end of the book that this ubiquity of consumption includes themselves.

7 Mayle continues this theme in *Hotel Pastis* where the protagonist manages to lose weight apparently without effort as a result of his new way of life. Even more impressively, he manages to spend evenings eating and drinking a great deal to awake the following morning when 'to his surprise he didn't have a hangover' (*Hotel Pastis*, 60).

8 Rumours of the possible return of the Mayles to Provence were claimed to have made 'natives jittery' (Harlow and Bodkin 1996). However, this time Mayle was determined to keep a low profile.

REFERENCES

Aldridge, A. (1995) 'The English as they see others: England revealed in Provence', *The Sociological Review* 43(3), pp. 415–34.

Blume, M. (1992) *Côte d'Azure: Inventing the French Riviera*, London: Thames and Hudson.

Blunt, A. and Rose, G. (eds) (1994) *Writing Women and Space: Colonial and post-colonial geographies*, New York: Guilford.

Bourdieu, P. (1984) *Distinction: A social critique of the judgment of taste*, London: Routledge.

Buller, H. and Hoggart, K. (1995) *International Counterurbanization: British migrants in rural France*, Aldershot: Avebury.

Fortescue, Lady (1935[1992]) *Perfume from Provence*, London: Transworld Publishers.

Fraser, N. (1996) 'Mad frogs and English phlegm', *Guardian*, 18 May, p. 25.

Gates, J. (1992) 'Go South to the sun to beat the recession', *Daily Express*, 5 August, p. 33.

Gregory, D. (1994) *Geographical Imaginations*, Oxford: Blackwell.

Harlow, J. and Bodkin, W. (1996) 'Mon Dieu! Mayle's return to Provence makes natives jittery', *The Sunday Times* (electronic version), 15 September.

Lefort, J.-F. (1993) 'An unforgivable success', *Times Literary Supplement*, 12 March, p. 14.

Lowenthal, D. (1985) *The Past is a Foreign Country*, Cambridge: Cambridge University Press.

McClure, J. (1994) *Late Imperial Romance*, London: Verso.

Martin, M. (1992) *A Year near Proxima Centauri*, London: Corgi Books.

Mayle, P. (1989) *A Year in Provence*, London: Pan Books.

—— (1991) *Toujours Provence*, London: Pan Books.

—— (1993) *Hotel Pastis: A novel of Provence*, New York: Alfred A. Knopf.

Ó Tuathail, G. (1994) '(Dis)placing geopolitics: writing on the maps of global politics', *Environment and Planning D: Society and space* 12, pp. 525–46.

Pratt, M.L. (1992) *Imperial Eyes: Travel writing and transculturation*, London: Routledge.

Sahlins, M. (1974) *Stone Age Economics*, London: Tavistock Publications.

Said, E. (1979) *Orientalism*, London: Vintage Books.

Thorpe, A. (1991) 'All that sunshine', *Times Literary Supplement*, 26 April, p. 21.

Urry, J. (1990) *The Tourist Gaze: Leisure and travel in contemporary societies*, London: Sage.

Wiener, M. (1981) *English Culture and the Decline of the Industrial Spirit, 1850–1980*, Cambridge: Cambridge University Press.

Williams, R. (1973) *The Country and the City*, London: Verso.

Winterbotham, H. (1989) 'Lure of the Lubéron', *Times Literary Supplement*, 4 August, p. 844.

Index

Note: Page numbers in **bold** type refer to **figures**; page numbers followed by 'N' refer to notes.

Abrams, M.H. 559
Adderly, C. 176
Ade, G. 137
Africa 8, 14–45, 38N; European contact 29; image 18; interior 16; maps 16, 17; pre-abolition narratives 37N
African trade 38N
AIDS 188
Aldrich, R. 89N
Alexandria 120, 127, 131
Ali, Mohammed 133
Allahabad 95, 96, 104, 105, 106, 108
Almond, P. 54, 56, 68N
Alpers, S. 71
Altman, D. 188
ambush: Greece (1870) 173
Andrews, M. 152
Angola 27
antiquities: exhibition 133–4
Appiah, A. 1–2
Appleton, T.G. 114, 136, 140, 147N
Arabian desert 81
Arabian Nights 139–40, 142–3, 144
Archer, W.G. 89N
Arnold, J. 124, 134
Arnold, M. 175
Assad, T. 87N
Athenaeum 95
Athens 176
Atkins, J. 18, 30, 39N, 40N, 42N, 44N

Bacon, L. 136
Baedeker, K.: *Egypt, Handbook for Travellers* 118, 129, 134–5, 141, **142**, 144; *Greece, Handbook for Travellers* 166, 170, 174, 175, 177

Baker, Sir S. 154–6
Barker, A. 15, 34, 38
Barret-Ducrocq, F. 88N
Barrow, J. 19
Bartlett, W.H. 114, 118, 120, 121, 123, 139, 148N
Bartrum, K. 97, 98, 101, 102, 106, 107, 110N
Batten, C. 6
Behdad, A. 8, 135
Bell, C.F.M. 140
Bell, D. 70
Belle, H. 164–5, 167
Bent, J.T. 166
bestiality 82
Bevan, E.R. 167–8
Bhabha, H. 9, 156–7
bisexuality 89N
Bishop, P. 66
Black Prince 34–5
blackness 8
blacks: British contact 16
Blume, M. 211
Blunt, W.S. 109
Bogle, G. 51–3, 58–9
Bogue, R. 190
Book of a Thousand Nights and a Night (Burton) 71, 72, 73, 75–83, 85, 87N, 88N
Boone, J. 76
Boothby, R. 26
border literature 167
Bosman, W. 19, 40N
boundaries 84
Bourdieu, P. 216
Bradley, R. 21–22
brigandage 173, 178

Brighton Pavilion 153
British Nabobs 152
British School 169
Brodie, F.M. 87N
Brooks, P. 186
Brown, L. 8, 18, 28, 30, 43N
Browne, Sir T. 21
Brydon, C. 100, 110N
Buddhism 50, 51, 54, 57, 68N
Buller, H. and Hoggart, K. 206, 214, 215
Burton, I. 74, 77, 84, 88N
Burton, R.F. 8, 70, 70–91, 87N
Butler, E. 121, 131, 135
Buzard, J. 116

Cairo 78, 120, 127–9, **128**, 131, 133, 139, 142–3
Calcutta 95, 106, 107, 108
Campbell, J. 80, 153
Campbell, Sir C. 95, 110N
campers 212
cannibalism 17, 39N, 43N
capitalism 193
Captain Singleton (Defoe) 8, 15, 25, 26, 28, 29, 30, 32, 34, 42N, 44N
Cardinal, R. 6
Carey, M.L.M. 126
caricatures: Greek 179–80
Carlyle, T. 60
Case, A. 97, 98, 100, 101, 103, 110N
Cave, H.W. 156, 157, 158, 159–60
Cawnpore 96, 102, 108, 111N; entrenchments **103**, 104
Ceylon 9, 151–63
Christianity 22, 43N
civility 29
Clarence, L.B. 161
class 108; difference 105; divide 111N
Clayton, D. 3
Clifford, H. 180N
Clifford, J. 164, 168–9, 197
closet 185–99; metaphor 187–8
cloth 44N
Cobbe, F.P. 125, 137, 141
Cochrane, G. 171–2
coins 43N
Colonial Encounters (Hulme) 17
colonialism 181N
colonialist hybridity 153
commodities 18
Compleat Geographer, The 14, 20

complexion 21
Congo Journey (O'Hanlon) 2
consumption 210
Cook, Captain J. 3, 4
Cook, T. 7, 118
costume 126, 129, 142
Côte d'Azur 211–12, 215
Criminal Law Amendment Act (1885) 83
Cross-Channel 216
cultural difference 19; trade 22
Curtin, P. 37N
Curtis, G.W. 117, 118, 126, 139, 141
Curtis, W. 131, 132–3, 134
Curzon, G. 167
Curzon, R. 123

dahabeeah 117, 121, **122**, 126
Dahomey 22, 24
Daily News 95
Daily Tribune 96
Dala Lama: Thirteenth 62
danger 143
Dashwood, A.F. 110N
Davies, K.G. 27, 38N
De Butts, L. 155
de Körös, A.C. 55, 68N
de Lamartine, A. 164
decay: social and physical 158–9
Defoe, D. 8, 15, 25–31, 34, 44N
Deleuze, G. and Guattari, F. 71, 190, 193–5
demonic representations 57
Dervishes 144
Deschamps, G. 165, 173
Description de l'Egypte 145
desire 74; lesbian and gay men 185–99
D'Estournelles, Baron P. 173
deterritorialization 193
Dilkusha Palace 111N
Dinapore 106
dis-orientation 151–63, 168
discourses: altered 33
domestic and imperial subjectivity 106, 107, 108
domestic space 94
Don Pacific incident (1847) 176
Douglas, J. 3
dragomans 120, 123, 130, 146N
Dragoumis, J. 180
Du Camp, M. 4
Duncan, J. and Duncan, N. 78
Dunning, H. 136

Eagleton, T. 186
Eames, J.A. 120, 135, 144
Earle, P. 26
Ebers, G. 127
Eden, F. 148N
educational curricula 38N
Edwards, A. 115, 118, 121, 122–3, 124, 126, 136, 140–1
Egypt 9, 114–48; 'new' 146; theatre metaphor 115–16
Egyptians 176
Ellis, H. 74
Engels, F. 96
English (people) 174, 175
Enlightenment: Greek 180–1N
Enzensburger, H.M. 6
epiphany: meanings 59; textualization 64
erotica: printing 82
Europe: agency/disassociating from slave trade 34; aggression 36; concept of value 29, 30; split behaviour 36
European 38N
evacuation 92, 99
Evangelism 59
excavation 132

Fabian, J. 8
Farrer, R.R. 165, 176–7
fashion 212, 216
Ferguson, R. 134
Ferriman, Z.D. 174
firearms 31
First Footsteps in East Africa (Burton) 78, 79, 87N
flagellation 88N
Flaubert, G. 3–4
Flight From Lucknow (Solomon) 92, **93**, 94
Forster, E.M. 164
Fortescue, Lady 203, 216
Foucault, M. 87N
four-stages theory 20, 41N
France: overseas colonies 166
Frazer, J. 169
Freud, S. 186
Frye, N. 79

Gallop, J. 186
gay men 10
gay oppression 185
Gemeinschaft 202, 208–9
gender power 94
geographies: of desire 189–95

George, R.M. 98
Germon, M. 96, 100, 101, 102, 104, 110N, 111N
Ghawazi 143
Gibson, C.D. 119, 145
globalization 213
gold 26, 27, 30
Gordon, Lady D. 127, 146N
Gourgouis, S. 181N
Government of India Acts (1858) 108
Grand Tour 6, 203
Greece 9–10, 164–84; classical and modern 169–70, 171; and Ireland 176–8
Greeks 173–6; English 175
Gregory, W. 153
Grenfell, W.T. 114
Guattari, F. and Deleuze, G. 71, 190, 193–5
guidebooks 134, 136, 139, 143, 168, 170, 214–15 *see also* Baedeker, K.; Murray, J.
gun-boat diplomacy 176

Haeckel, E. 160–1
Halbwachs, M. 4
Hale, M. 41N
Harris, K. 97, 98, 99, 102, 104, 105, 110N, 111N, 112N
Hastings, W. 54, 67N
Hellenism 167, 169–70
Herzfeld, M. 180N
Hinduism 51
Hocquenghem, G. 190, 193–4
Hodgson, B. 54–6, 67N
Hoggart, K. and Buller, H. 206, 214, 215
Homer 172
homophobia 83, 84; Victorian 83
homosexuality 74, 81, 83, 87N, 88N
Hong Kong 192, 194
Hopley, H. 118
Hoskins, G.A. 121, 124
Hotel Pastis (Mayle) 201–2, 213–14, 217N
hotels: Alexandria 120; Cairo 120, **128**, 128–9
Houstoun, J. 18
Hulme, P. 17
human difference: origin 19–20
human sacrifice 42N
humanism 53

Huxham, A.E. 99, 107, 110N, 111N
Hyam, R. 77

Ibrahim the Copt 123
Illustrated London News 95, 109N
imagery: sexual 82, 83
imaginary geography 9, 151
imperial crisis: domestic terms 108
imperial travel writing 9
imperial/domestic subjectivity 101
imperialism 64, 94, 171; acquisition
    18
India: British women 92, 94, 109;
    mutiny/uprising (1857–9) 92
Indian taste 152
industrialized romanticism 7
Inglis, J. 96, 99, 100, 101, 102, 103–4,
    105–7, 110N, 111N, 112N
Innes, M. 110N
interior: Africa 16
Into the Heart of Borneo (Appiah) 1–2
Ireland: and Greece 176–8
iron 43N
Islam 144
ivory 26, 27

Jablow, A. 38N
Jebb, R.C. 166, 168, 177
Journal of a Voyage up the Gambia 22

Kabbani, R. 86, 89N
Kama-shastra Society 82
Kama Sutra 82
Kandy (Ceylon) 153–4, **154**, 156, 158
Kandyan Highlands (Ceylon) 151–63
Kawaguchi, E. 62
Kaye, Sir J. 98
Kelly, R.T. 137, 142
Khedive Ishmail 127
Kipling, R. 57–9, 60–1, 68N, 137–8;
    Kim 57–9, 60–1

Lacan, J. 186, 189–90, 190–3, 196
Lady's Newspaper 107, 110N
Lake Regions, The (Burton) 79
Lamaism 56, 62
Lane, E.W. 140, 146
Laporte, L. 115
Laurens, H. *et al.* 145
Law of Arms 34
Le Vice contre nature 84
Leland, C. 114, 127
Lewis, F. 161
Lhasa 62

literary theory: and psychoanalysis
    186–7
Lobsang Palden Yeshe (Third Panchen
    Lama) 52, 67N
Lorimer, N. 129, 130, 137
Low, S. 138–9
Lucknow 9, 92–112
Lucknow Relief Committee 107, 108
Luke, T. 189
Luxor 123, 129
Lynch, E.M. 177

McClintock, A. 94
McClure, J. 215–16
McLynn, F. 88N, 89N
Madagascar 26–27, 28
Mahaffy, J.P. 166, 177–8
*Manners and Customs of the Modern
    Egyptians* (Lane) 140
mapping 71
Marcus, S. 82, 88N
Marden, P.S. 129
Mariette, A. 132
Markham, C. 53, 58
Marshall, P.J. and Williams, G. 39N
Martin, M. 217M
Martineau, H. 115, 120, 126, 132,
    147–8N
masculinity: Provençal 207
Mason, M. 83, 84, 88N, 89N
Massey, D. 84
Matheson, J. 156
Maule, F. 118
Mayle, P. 10, 200–18
mercantilism 53
Metcalf, T. 109N
migration: to rural France 206, 215
Miller, N. 185, 188–9, 191–2, 194–5,
    196
mimic place 9
Minh-Ha, T. 153
Mitchell, T. 115
monks: Tibetan 62
monogenesis 19
Montbard, G. 136
Moore, J. 20
Morgan, P. 35
Morris, J. 78
mortality 16
Moussa-Mahmoud, F. 78
Murray, J.: *Handbook for Travellers to
    Egypt* 132, 135, 139, 141; *Handbook
    for Travellers to Greece* 169, 170,
    172–3, 173–4, 174–5, 177, 179

museum: Lahore 57
Mycenaean culture 178
Myth of Shangri-La (Bishop) 66

nakedness 22, 32, 41N, 44N
national flag 147N
National Vigilance Association 83
natural resources 18
Negro image 39N
Nerval 7
Newara Eliya 154–5, **160**, 160
Nicolson affair (1885) 176
Nightingale, F. 137
Nights (Burton) *see Book of a Thousand Nights and a Night* (Burton)
Nordau, M. 166
Novak, M. 42N

Ó Tuathail, G. 216N
O'Hanlon, R. 1, 2
Oedipus complex 193
Orientalism 54, 115, 119, 145, 153, 171, 214; imaginary 139
ornamentation 43N
Osysseus 179
Ottoman Empire 171

Pall Mall Gazette 83
Palmerston, Lord H.J.T. 176
Parisians 211, 215
pastis 205
Pearsall, R. 83
pederasty 73, 74, 75, 84, 87N; types 82
Penfield, F.C. 131
Perfume from Provence (Fortescue) 216
Personal Narrative (Burton) 81–2
Petrie, F. 119
phantasmagoria 139–40
Philosophical Transactions of the Royal Society 21
photography 4, 145
Pictorial Times 107, 110N
Pilgrimage to El-Medinah and Meccah (Burton) 86N
pirates 27
pleasure 6
political criticism 83
polygenesis 20
pornography 82, 83
pornotopia 82, 83, 84
Porter, D. 6, 7
post-coloniality 180N; studies 171–2

post-Fordism 206
power politics 24
Pratt, M.L. 19, 39N 210
Prime, W. 117, 120, 121, 123
primroses 157–8
prisoners 33–4, 44N
Proust, A. 173
Provence 200–18; 'industry' 201; as pastoral England 206–8
psychoanalysis 193, 197N; and literary theory 186–7

Qualls, B.V. 68N

racial characteristics 20
Ramsay, W.M. 169–70
reading 117
reciprocity concept 210
Red Caps 51
Reeve, C.M. 126, 148N
religion: Tibetan 50, 51–3, 54
representation: spaces 3–5
Reynolds, Sir J. 152
Reynolds-Ball, E. 127, 133, 139, 144
Richetti, J. 34
Rinpoche, G.T. 62
Roberts, E. 120
Rodney, W. 31
Rogers, K.R. 158
Romans 176
romanticism 6, 151, 152, 153, 155, 158, 215; industrialized 7
Royal Academy 92
Royal African Company 37N, 39N, 44N
Royal Proclamation (1858) 108
ruins 148N
Russell, W.H. 96, 102

Said, E. 5, 8, 59, 86, 89N, 115, 167, 185, 214; on desert 81; textual attitude 56, 116, 171
Sanskrit 55
Saturday Review 85
savagery 17, 31, 32
savants 114
schizoanalysis 193, 194, 195, 196
Schliemann, H. 172
Scrimgeour, G. 28
scripting 116–17, 145–6
Seaver, G. 62, 65, 69N
Secord, A. 25
secular revelation 59–61
Sedgwick, E. 185
Selden, R. and Widdowson, P. 186

self-realization 66
Selma 191–2, 194
Sergeant, L. 167, 172, 174, 175
servitude 98, 99
sexual geography 85
sexual(ity) 70–89; attitudes 84; discourse
   81; imagery 82, 83; Victorian 8–9
Sharpe, J. 98, 111N
sights 134
sightseeing 117, 136
Sikandar Bagh 100, 111N
skin colour 21
Sladen, D. 117, 129, 130, 134, 143,
   145
slavery 22–3, 44N; abolition 45N; trade
   30, 33,
34, 36, 37N
Smith, A. 168, 173
Smith, A.C. 124
Smith, B. 4
Smith, N. 71
Smith, W. 14
Smithers, L.C. 88N
Snelgrave, W. 14, 16, 17, 22, 24, 37N,
   40N
social integration 210–11
social memory 4
socio-economic equilibrium 210
Solomon, A. 92, **93**, 94
Something of Myself (Kipling) 60
Sotadic Zone 8–9, 70, 72–4, 77, 80–2,
   84–5
Speke, J.H. 72, 87N
spiritualism: secularized 63
spirituality: exploration 65; realization
   66
Spurr, D. 155
Stead, W.T. 83
Steevens, G.W. 165
Stephens, J.L. 147N
stereotypes 125, 145
superiority 20
Supplemental Nights (Burton) 83
Sutherland, J. 44N
Switzerland 156
Symonds, J.A. 74, 87N

Taylor, B. 121, 125, 126–7
Terminal Essay (Burton) 75, 83
Teshu Lama 53, 58
text 116
textual attitude (Said) 56, 116,
   171
textual enlightenment 51–4

textualization 64
Thackeray, W.M. 120, 169, 176
theatre 115–16
therapeutic discourse 200
Thomas Cook & Sons 129–30
Thorpe, A. 202
Tibet 8, 49–69
Times Literary Supplement (TLS)
   202
Times, The 95, 96, 169, 173
Tirard, H. and Tirard, N. 132
Todorova, M. 174
Toujours Provence (Mayle) 201, 202,
   204–5, 207–8, 211, 217N
tourism: effects 211
tourist sovereignty 126
Tourist, The 209
Townsend, M.W. 167, 176
Tozer, H.F. 175
trade 18, 30; cultural difference 22; slave
   30, 33, 36, 37N
trading 15–16
translation 4–5, 164
travel: modern 7; spaces 5–8; writing 9,
   200
travel guides 168 *see also* guidebooks
*Travels into the Interior of South Africa in
   the Years* (Barrow) 19
Trikoupis, C. 177, 178
Turner, S. 53, 67N
Tyndale, W. 144

unpublished accounts 110N
Upward, A. 169

veneration 52
Venuti, L. 5
vice 143
Victorian sexuality 8–9
view-hunting 137, 139
visual difference 19
Voyage en Orient (Nerval) 7

Waddell, L.A. 56–7, 62, 68N
Walker, K. 88N
Wallis Budge, E.A. 118
Warburton, E. 127, 145
Ward, J. 129
Warner, C.D. 114, 115, 124, 136
Warren, W.W. 121
Wells, F. 104, 106
Widdowson, P. and Selden, R.
   186
Wilde, O. 83

Wilkinson, J.G. 117–18, 120, 131–2, 134, 135, 144, 147N
Williams, G. and Marshall, P.J. 39N
Woodcock, G. 51
Woolf, B.S. 157
Woolf, V. (formerly Stephen) 175
Wright, E. 186

writing: anthropological style 203–6; colonialist style 203 *see also* scripting; text
Wyse, Sir T. 174

Year in Provence, A (Mayle) 10, 200–18
Yellow Caps 51
Younghusband, F. 49, 63–4, 68N